AWS B2.1-84

Standard for Welding Procedure and Performance Qualification

D1609260

Prepared by
AWS Committee on Qualification

Under direction of
AWS Technical Activities Committee

Approved by
the AWS Board of Directors

Effective date:
January 1, 1984

AMERICAN WELDING SOCIETY, INC.
550 N.W. LeJeune Road, Miami, FL 33126

International Standard Book Number: 0-87171-235-0

American Welding Society, 550 N.W. LeJeune Road, Miami, FL 33126

Note: By publication of this standard, the American Welding Society does not insure anyone utilizing the standard against liability arising from the use of such standard. A publication of a standard by the American Welding Society does not carry with it any right to make, use, or sell any patented items. Each prospective user should make an independent investigation.

This standard is subject to revision at any time by the responsible technical committee. It must be reviewed every five years and if not revised, it must be either reapproved or withdrawn. Comments (recommendations, additions, or deletions) and any pertinent data which may be of use in improving this standard are requested and should be addressed to AWS headquarters. Such comments will receive careful consideration by the responsible technical committee and you will be informed of the committee's response. Guests are invited to attend all meetings of AWS committees to express their comments verbally. Procedures for appeal of an adverse decision concerning your comments are provided in the Rules of Operation for AWS Technical Committees. A copy of these Rules can be obtained from the American Welding Society, 550 N.W. LeJeune Road, Miami, FL 33126.

Printed in the United States of America

Contents

Foreword

This Document provides requirements for welding procedure and welding performance qualification. It is the intent that this Document be referenced by other documents, such as codes, standards, specifications, or contracts.

This Document defines and establishes qualification variables. The criterion for determining if a welding condition is a qualification variable is whether or not a change in that condition beyond the allowable tolerance will affect the properties of a sound weld to the extent that the properties will not meet the specified minimum.

Qualification requirements are based on the premise that the referencing document will specify fabrication, design, base metal, filler metal, preheat, interpass temperature, postweld heat treatment, nondestructive examination, and test requirements applicable to the end product. Welding procedures and performance qualifications which meet the requirements of other documents are acceptable provided they also meet the requirements of this Document.

Welding Procedure Qualification

The purpose of a welding procedure qualification is to provide test data for assessing the properties of a welded joint. It is the obligation of manufacturers to produce welds which have properties suitable for the application. The proof of production weld soundness is determined by the type and extent of testing and examination applied, which is the responsibility of the referencing document.

This Document also provides for welding procedure qualification of special test weldments by performing tests simulating service conditions which may include impact loading, flexural loading, static loading, or cyclic loading to duplicate the type of loading which the weldment will encounter in service. The details of required tests and examination of special test weldments shall be specified by the referencing document.

Standard Welding Procedure Specifications published by AWS and as permitted by this Document (Section 2) may be used without performing additional procedure qualification tests when the requirements of the referencing document for the applications involved are met and no change is made beyond the specified range of variables shown in the welding procedure.

Welding Procedures Specifications other than Standard WPS published by AWS must be individually qualified by testing to demonstrate that minimum requirements are met.

Welding Performance Qualification

The purpose of welder qualification tests is to determine the ability of welders to deposit sound weld metal following a Welding Procedure Specification (WPS) and in the welding positions which will be encountered in production applications.

The purpose of welding operator qualification tests is to determine the ability of welding operators to operate machine or automatic welding equipment in accordance with a WPS.

Performance qualification shall be by mechanical or nondestructive testing of weldments, or both, except that provision is made for qualification by visual examination only when permitted by the referencing document.

Materials

Base metals and filler metals have been grouped into categories which will minimize the number of qualifications required. Substitution of one base metal or filler metal for another, even when within the allowable rules, should only be made after an evaluation of the factors involved. For some materials or combinations of materials, additional tests may be required. Materials not listed in Appendices B and C shall each require separate qualification (see 2.7.2 and 2.7.3).

Standard for Welding Procedure and Performance Qualification

1. General Provisions

1.1 Scope

This Document provides the requirements for the qualification of welding procedures.

It also provides the requirements for the performance qualification of welders and welding operators for manual, semi-automatic, machine, and automatic welding.

This Document is intended for use where referenced by a product code, standard, or specification. In cases where there is no referencing product document, this Document may be referenced for welding qualification in contract documents and the requirements for preheat, interpass temperature, postweld heat treatment, and fracture toughness shall be provided in those contract documents.

This Document is intended for use with the following welding processes:

OFW	=	Oxyfuel Gas Welding
SMAW	=	Shielded Metal Arc Welding
GTAW	=	Gas Tungsten Arc Welding
SAW	=	Submerged Arc Welding
GMAW	=	Gas Metal Arc Welding
FCAW	=	Flux Cored Arc Welding
PAW	=	Plasma Arc Welding
ESW	=	Electroslag Welding
EGW	=	Electrogas Welding
EBW	=	Electron Beam Welding
SW	=	Stud Welding

1.2 Terms and Definitions

1.2.1 The welding terms used in this Document shall be interpreted in accordance with the definitions given in the latest edition of AWS A3.0, *Welding Terms and Definitions.*

1.2.2 Supplemental Definitions. Some A3.0 terms, exceptions to A3.0 terms, and additional terms as referenced in this Document are defined below.

Employer - The term employer shall mean the contractor or manufacturer who produces the weldment for which welding procedure and performance qualifications are required. Whenever approval, signature, or certification by the employer are required by this Document, it shall mean the employer or a designated employee within his organization. Closely related companies, including those with different names for which effective control of welding is as one organization, shall be considered as one employer.

Filler Metal Number - Designated as F-number in this Document.

Pipe - Pipe is used generally to refer to pipe and tube.

Plate - Plate is used generally to refer to flat products and structural shapes.

Material Number - Designated as M-number in this Document.

1

Procedure Qualification Record (PQR) - A document providing the actual welding variables used to produce an acceptable test weld, and the results of tests conducted on the weld for the purpose of qualifying a welding procedure specification.

Qualifier - The term qualifier shall mean the employer or the organization, or individual specified by the referencing document as responsible for conducting and supervising the qualification testing.

Qualified Welding Procedure - A welding procedure meeting the qualification requirements of this Document based on qualification tests of weldments made in accordance with the WPS and recorded on a PQR. Qualified welding procedures are of two types:

(1) Procedures independently qualified by an employer for his own use.

(2) Standard procedures permitted for use without further qualification.

Qualified Welder - One who is qualified to the requirements of this Document to perform manual or semi-automatic welding.

Qualified Welding Operator - One who is qualified to the requirements of this Document to perform machine or automatic welding.

Qualification Variables - Those welding variables which, if changed beyond the limitations specified, require requalification of the procedure (see 2.7).

Test Weldment - Metal joined by welding for the purpose of qualifying welding procedures and/or welders or welding operators. The base metal thickness of the test weldment is commonly identified as "T" in this Document.

Test Specimen - That portion of a test weldment which is prepared for evaluation for qualification purposes. The test specimen thickness or weld metal thickness, as applicable, is commonly identified as "t" in this Document.

Welding Procedure Specification (WPS) - A document which delineates or references all welding variables required by this Document to provide direction for welding.

Welding Variable - Welding data which shall be recorded on the WPS and PQR (see 2.6).

1.3 Responsibilities

1.3.1 The employer shall be responsible for the welding performed by his organization, including the use of qualified welding procedures and qualified welders and/or welding operators.

1.3.2 The welding procedure may be a standard WPS, or it shall be qualified as required under the rules of Section 2.

1.3.3 It is the employer's responsibility to assure that Welding Procedure Specifications meet any additional requirements of the referencing document.

1.3.4 The employer shall be responsible for the application of welding procedures in production.

1.3.5 Records - Each employer shall maintain the applicable WPS's, PQR's, and Welding Performance Qualification Records during the period of their use.

1.4 Effective Date

When not otherwise specified by the referencing document, the edition of this Document to be used shall be established in accordance with the following:

(1) Editions may be used at any time after the effective date of issue.

(2) Editions become mandatory for new contracts six months after the effective date of issue..

(3) In cases where there is no specific contract date, the edition to be used shall be mutually agreed upon by the concerned parties.

(4) Editions established by contract date may be used during the entire term of the contract, or the provisions of later editions may be used when agreed upon by the contracting parties.

1.5 Safety Precautions

Adequate safety precautions, including ventilation of the work area, shall be taken in accordance with the requirements of ANSI Z49.1, *Safety in Welding and Cutting.**

---

*Available from American Welding Society, 550 N.W. LeJeune Road, Miami, FL 33126.

2. Welding Procedure Qualification

2.1 General

2.1.1 Two categories of qualified welding procedures are established in this Section. (See Fig. 2.1.1.)
    (1) Procedures qualified (2.2) using
        (a) Standard Qualified Weldments, or
        (b) Special Test Weldments
        and
    (2) Standard WPS (2.3)

2.1.2 Each welding procedure shall be qualified to establish the properties which are expected to result from its application to production weldments. The primary purpose of welding procedure qualification is to establish the properties of the test weldment.

2.1.3 A matrix indicating welding variables to be included in the WPS and documented on a PQR for each process or combination of processes is provided in 2.6. Suggested WPS/PQR forms are illustrated in Appendix A, Forms A.7.1 and A7.2. Other formats may be used providing all applicable information is recorded, including the certifying statement shown on the suggested form.

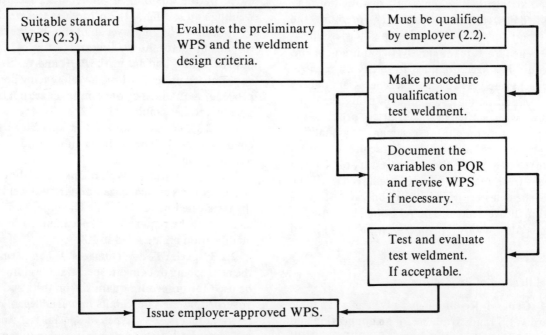

Fig. 2.1.1 — Steps to determine procedure qualification requirements

3

2.1.4 PQR's shall not be revised except to correct errors or add omitted information; however, all such changes shall be identified and dated on the PQR.

2.1.5 A WPS may require the support of more than one PQR, while one PQR may support a number of WPS's. PQR's qualified using different processes may be combined to support a WPS.

2.1.6 WPS's and PQR's shall be identified according to a system which allows permanent traceability from the WPS to its supporting PQR's. Each WPS shall be uniquely identified.

2.1.7 A change in any qualification variable beyond ranges allowed in 2.7 shall require requalification of the WPS and preparation of a new or revised WPS. Other changes shall not require requalification provided such change is documented either in a new, revised, or amended WPS.

2.1.8 Base metals not listed in Appendix C shall require qualification testing of each individual WPS to be used with that base metal. Welding procedures for unlisted base metals shall not be used for welding listed base metals or vice versa.

2.1.8.1 Coated metals, such as galvanized or painted metals, shall require separate qualification if the coating is not removed from the weld area prior to welding, unless otherwise permitted by the referencing document.

2.1.8.2 Weld Surfacing. Weld cladding and hardfacing require separate qualification for each base metal M-number and filler metal combination [2.7.3(2)]. Welds made to join clad metals to other clad metals or to unclad metals shall be separately qualified or may be qualified by a combination of a WPS for joining the unclad metal and a WPS for applying the cladding.

2.1.9 When fracture toughness is a requirement and a procedure has been qualified to satisfy all requirements except fracture toughness, it is necessary only to prepare an additional test weldment using the same procedure with sufficient material to provide the necessary fracture toughness specimens. If a qualified welding procedure has satisfactory fracture toughness values in the weld metal, then it is necessary only to test specimens from the heat-affected zone (HAZ). If the HAZ has been tested, it is necessary only to test specimens from the weld metal. The WPS shall be revised to accommodate fracture toughness variables (T) listed in 2.7.

2.2 Qualification of Procedures

2.2.1 General Requirements. Where a suitable Standard WPS is not available for an application and if the employer does not have a qualified WPS, a WPS shall be prepared and qualified.

2.2.1.1 Welders and welding operators shall be under the full supervision and control of the qualifier during the welding of test weldments. The qualifier is responsible for assuring proper:

(1) Preparation of test materials for welding.

(2) Documentation of welding procedure (see 2.2.1.2 and 2.2.1.3).

(3) Preparation of test specimens from the completed weldment.

(4) Performance of examinations and mechanical tests.

(5) Documentation of test results.

(6) Welding procedure instruction is given to the welder or welding operator.

2.2.1.2 To qualify a welding procedure, a preliminary Welding Procedure Specification shall be used to make a test weldment. All welding information required to make the test weldment shall be shown in this WPS. Any changes made in the welding variables prior to or during qualification shall be recorded in this WPS.

2.2.1.3 The WPS identification plus the actual ranges and details of welding variables used in making the test weldment shall be recorded on a PQR.

2.2.1.4 The completed test weldment shall be subjected to the tests and examinations required for either standard (see 2.2.2) or special (see 2.2.3) test weldments, and the results shall be recorded on the PQR.

2.2.1.5 If the results meet the acceptance requirements specified in 2.5, the PQR shall be signed and dated by the qualifier as an accurate record of the welding and testing of the procedure test weldment.

2.2.1.6 The employer shall signify acceptance of responsibility for the welding and testing of the procedure by signing and dating the PQR and WPS. Except as permitted under 2.3.1 or the referencing document, procedures qualified by one employer are not transferable to another employer.

2.2.1.7 Preparation of test specimens for mechanical tests shall be as shown in Appendix A of this Specification.

2.2.2 Standard Test Weldments. Procedure qualification requires evaluating a standard test weldment by the tests listed in Table 2.2.2. The type, number, and location of the required test specimens for procedure qualification are detailed in 2.8.

2.2.3 Special Test Weldments. When permitted by the referencing document, special test weldments may be used for procedure qualification tests and shall be governed by the same limits on variables as standard test weldments. Two types of special test weldments are recognized by this document, and requirements for testing each are outlined below.

Table 2.2.2
Test requirements for standard test weldments

Type of test	Groove welds	Fillet welds	Stud welds	Re-bar	Weld cladding	Hard facing
Visual examination	Yes	Yes	Optional	Optional	Optional	Optional
Guided bend tests	Yes	—	—	—	Yes	—
Tension test	Yes	—	Yes or torque	Yes	—	—
Macro-examination	—	Yes	Yes*	Yes	—	Yes
Bend tests	—	—	Yes	—	—	—
Torque tests	—	—	Yes, or tension	—	—	—
Fracture toughness tests	If specified	—	—	—	—	—
Shear test	—	Yes	—	—	—	—
Penetrant examination	—	—	—	—	Yes	If specified
Chemical analysis	—	—	—	—	Yes	If specified
Hardness test	—	—	—	—	—	Yes

*Except that unclad, unpainted M1 materials are exempted.

2.2.3.1 Simulated Service Test Weldments. These are test weldments in which qualification shall require making tests simulating service conditions. These tests may include impact loading, flexural loading, static loading, or cyclic loading to duplicate the type of loading which the weldment will encounter in service.

2.2.3.2 Prototype Structure Test Weldments. These are test weldments in which qualification requires that a prototype of the actual weldment be subjected to field tests in which it is loaded and demonstrated to perform the function for which it was designed.

2.2.3.3 Required Tests and Examination. The details of required tests and examination of special test weldments shall be specified by the referencing document or contract agreement. Where any test in Table 2.2.2 is specified, the acceptance criteria shall be as required in 2.5. The acceptance criteria for other tests shall be as specified by the referencing document or contract agreement.

2.3 Standard Procedures

2.3.1 Procedures allowed for use without qualification tests shall be limited to those known as Standard WPS's. WPS's may be copied without specific authorization as outlined below.

2.3.2 Employers may use a Standard WPS without modification if it provides sufficient direction for use in production. An employer may supplement Stand- and Procedures with additional information so long as the qualification variables remain within the allowable ranges of the Standard WPS.

2.3.3 The employer may prepare a specific WPS based on a Standard WPS provided the qualification variables fall within the allowable ranges of the Standard WPS. The applicable Standard WPS shall be attached to or referenced on the employer's specific WPS.

2.3.4 Prior to use of a Standard WPS, the employer shall signify acceptance of responsibility for the application of the procedure by signing and dating the Standard WPS, or the employer's specific WPS, whichever is used for production welding.

2.3.5 Any change in a qualification variable beyond the ranges allowed in 2.7 shall require a new or revised WPS. Where a Standard WPS cannot be found to cover the change, a new or revised WPS shall be qualified under the rules of 2.2. Changes in variables other than qualification variables shall be documented by an amendment attached to the applicable WPS or by revision of the WPS.

2.4 Qualification Limitations

2.4.1 Thickness Ranges. Tables 2.4.1 A, B, C, and D specify limitations on thickness ranges qualified, based upon the weld metal thickness of a complete joint penetration groove test weldment deposit thickness, or fillet test weldment leg size, or base metal thickness for weld cladding and hardfacing, or weld metal thickness of reinforcing bars.

2.4.1.1 If not otherwise stated in the referencing documents, successful procedure qualification of complete joint penetration groove weld shall qualify: (1) partial joint penetration groove welds based on the thickness of weld metal in accordance with Table 2.4.1A, and (2) fillet welds based on the thickness of weld metal in accordance with Table 2.4.1C.

2.4.1.2 When multiple process or multiple filler metal classifications are used in a single test weldment, the thickness ranges apply separately to each welding process and filler metal classification, and shall be documented on the PQR.

Table 2.4.1A
Qualification thickness limitations for groove welds[7]*

Thickness T of test weldment, in.	Range of thickness T of base metal qualified, in.		Range of thickness t of weld metal qualified, in.	
	Min	Max	Min	Max
Less than 1/16	T	2T	t	2t
1/16 to 3/8 incl.	1/16	2T	1/16	2t
Over 3/8, but less than 3/4	3/16	2T	3/16	2t
3/4 to less than 1-1/2	3/16	2T	3/16	2t when t $<$ 3/4 2T when t $\geq$ 3/4
1-1/2 and over	3/16	8	3/16	2t when t $<$ 3/4 8 when t $\geq$ 3/4
3 and over	1	Unlimited	1/2	2t when t $<$ 3/4 Unlimited when t $\geq$ 3/4

Thickness T of test weldment, mm	Range of thickness T of base metal qualified, mm		Range of thickness t of weld metal qualified, mm	
	Min	Max	Min	Max
Less than 1.6	T	2T	t	2t
1.6 to 9.5 incl.	1.6	2T	1.6	2t
Over 9.5, but less than 19	4.8	2T	4.8	2t
19 to less than 38	4.8	2T	4.8	2t when t $<$ 19 2T when t $\geq$ 19
38 and over	4.8	203	4.8	2t when t $<$ 19 203 when t $\geq$ 19
76 and over	25.4	Unlimited	12.7	2t when t $<$ 19 Unlimited when t $\geq$ 19

*See Notes 1, 2, 3, 4, 5, 6, 7, and 8 at the end of Table 2.4.1D.

Table 2.4.1B
Qualification thickness limitations for reinforcing bar[3]*

Type of test weldment	Weld metal thickness t range qualified - reinforcing bar					
	Groove				Fillet	
	Butt		Spliced			
	Min	Max	Min	Max	Min	Max
Butt joint	Any	t	Not qualified		Any	Any
Spliced butt joint (flare groove)	Not qualified		Any	t	Any	Any
Fillet test weld	Not qualified		Not qualified		Any	Any

*See Notes 3 and 4 at the end of Table 2.4.1D.

Table 2.4.1C
Qualification thickness limitations for fillet welds[8]*

Fillet test weldment	Size and thickness range qualified	
	Base metal thickness	Fillet size (leg)
Single pass	Unlimited	Maximum welded single pass fillet size and smaller
Multiple pass	Unlimited	1/2 from that welded to unlimited

*See Notes 3, 4, 5, 7, and 8 at the end of Table 2.4.1D.

2.5 Acceptance Criteria

Procedure qualification test specimens shall meet the following acceptance criteria for the tests required in 2.2.2 and 2.2.3.

2.5.1 Visual Examination. Visual examination is the examination of the test weldment without magnification other than a corrective lens. The test weldment may be examined by the qualifier at any time and the test may be terminated at any stage for modification of the preliminary Welding Procedure Specification. Acceptance criteria for visual examination are:

(1) The weld face shall exhibit uniformity, shall merge smoothly into the base metal, and have no abnormal roughness or other discontinuities.

(2) There shall be no cracks or incomplete fusion.

(3) There shall be no incomplete joint penetration in groove welds.

(4) Undercut shall not exceed the lesser of 10 percent of the base metal thickness or 1/32 in. (0.8 mm).

2.5.2 Bend Tests

(1) Groove Welds - For transverse bend specimens, the weld metal and HAZ shall be completely within the bent portion of the specimen after testing.

(2) Guided Bend Specimens - Guided-bend specimens shall have no open defect exceeding 1/8 in. (3.2 mm), measured in any direction on the convex surface of the specimen after bending, except that cracks occurring on the corners of the specimen during testing shall not be considered, unless there is definite evidence that they result from slag inclusions or other internal defects.

(3) Weld Cladding - For corrosion-resistant cladding, no open defect exceeding 1/16 in. (1.6 mm) measured in any direction on the surface shall be per-

Table 2.4.1D
Qualification thickness limitations for cladding and hardfacing[9]*

Test weldment base metal thick (T), in. (mm)	Thickness of qualified base metal, in. (mm)	
	Min	Max
Less than 1 (25.4)	T	Any
1 (25.4) and over	1 (25.4)	Any

*See Note 9.

Notes:

1. ESW, EGW, EBW shall only qualify within plus or minus 10 percent of the test weldment deposit thickness.

2. When the groove is filled using a combination of welding processes and/or welding procedures with the same process (one welding process with a different combination of qualification variables):

   a. The thickness t of the weld metal for each welding process shall be determined and used in the Range of Thickness t of Deposited Weld Metal column. The test weldment thickness T is applicable for the base metal for each welding procedure, and shall be determined and used in the Range of Thickness T of Base Metal column.

   b. When the weld metal thickness by any welding process or any welding procedure is 3/8 in. (9.5 mm) or less, this minimum shall be 1/16 in. (1.6 mm) for that welding process.

   c. Each welding process and each welding procedure qualified in this combination manner may be used separately only within the same qualification variables and within the thickness limits described in Notes 3 through 9.

3. For OFW, the maximum base metal thickness qualified is the thickness of the test weldment, and the maximum weld metal thickness qualified is the thickness of the weld metal of the test weldment.

4. For the short-circuiting transfer mode of GMAW, the maximum base metal thickness qualified shall be 1.1 times the test weldment thickness, and the maximum weld metal thickness qualified shall be 1.1 times the weld metal of the test weldment.

5. For fracture toughness applications less than 5/8 in. (15.9 mm) thick, the base metal thickness of the test weldment is the minimum base metal thickness qualified.

6. Where any single pass weld is greater than 1/2 in. (13.7 mm) in thickness, the qualified base metal thickness is 1.1 times the test weldment thickness.

7. If the test weldment receives a postweld heat treatment exceeding the lower critical temperature, the maximum base metal thickness qualified is 1.1 times the base metal thickness of the test weldment, and the maximum weld metal thickness qualified is 1.1 times the weld metal of the test weldment.

8. For M-11 steels, the fillet size shall be equal to or less than the fillet size qualified.

9. The minimum base metal thickness qualified for cladding and hardfacing is one layer if the test weldment has only one layer, and is two layers if the test weldment has two or more layers. The number of layers applies individually to each filler metal.

mitted in the cladding, and no open defects exceeding 1/8 in. (3.2 mm) shall be permitted at the weld interface after bending.

(4) Stud Welds - Studs shall be bent 15 degrees using a jig similar to that shown in Appendix A, Figure A6.2A, or 15 degrees by hammer blows and returned to essentially the original position. To pass the test(s), each of five welded studs and heat-affected zones shall be free of visible separation or fracture after bending and return bending.

2.5.3 Tension Tests

(1) Plate or Pipe Groove Welds - To pass the tension test, each test specimen shall have a tensile strength not less than:

(a) The specified minimum tensile strength of the base metal, or

(b) The specified minimum tensile strength of the weaker of the two base metals if base metals of different minimum tensile strength are used, or

(c) The specified minimum tensile strength of the weld metal when the applicable design provides for the use of weld metal having lower tensile strength at room temperature than the base metal, or

(d) If the specimen breaks in the base metal outside of the weld or weld interface, the test shall be accepted as meeting the requirements, provided the strength is not more than 5 percent below the specified minimum tensile strength of the base metal.

(2) Stud Welds - The failure strength shall be based on the minor diameter of the threaded section of externally threaded studs, except where the shank diameter is less than the minor diameter, or on the original cross-sectional area where failure occurs in a non-threaded, internally threaded, or reduced-diameter stud. To pass the tension test, each of five welded studs shall have a tensile strength as shown below:

(a) For ferrous metals, not less than 55,000 psi (380 MPa).

(b) For nonferrous metals, not less than 1/2 of the tensile strength of the base metal.

(3) Reinforcing Bar Welds - To pass the tension test, each specimen shall have a tensile strength that is not less than:

(a) 125 percent of the minimum specified yield strength of the type and grade of bar being joined (or the bar having the lower minimum specified yield strength for dissimilar joints), unless otherwise specified by the referencing document or by the qualifier if there is no referencing document.

(b) For welds between reinforcing bars and structural shapes or plate, the lesser of the specified minimum tensile strength of the structural shape or plate material or 125 percent of the minimum specified yield strength of the type and grade of bar being joined, unless otherwise specified by the referencing document, or by the qualifier if there is no referencing document.

2.5.4 Chemical Analysis of Cladding and Hardfacing Welds. The results of the required chemical analysis shall be within the range of analysis specified in the WPS.

2.5.5 Macro-Examinations - (Appendix E)

2.5.5.1 Macroetch cross-sections shall be polished and etched to provide a clear definition of the weld metal and heat-affected zone. Visual examination of etched surfaces shall be without magnification unless otherwise specified below:

(1) Stud Welds - To pass the macro-examination, each of five sectioned welds and the heat-affected zones shall be free of cracks when examined at a magnification of 10X.

(2) Hardfacing Welds - To pass the macro-examination, each of two faces of the hardfacing exposed by sectioning shall be examined with 5X magnification for discontinuities in the weld and base metal and HAZ, and shall meet the acceptance requirements specified in the referencing document.

(3) Fillet Welds - To pass the macro-examination, cross-sections of the weld metal and heat-affected zone shall show complete fusion, freedom from cracks, and no more than 1/8 in. (3.2 mm) difference in the length of the fillet legs.

(4) Rebar Welds

(a) For single-V or double-V-groove welds, thermit welds, and pressure gas welds, to pass the macro-examination, each specimen shall have complete joint penetration and complete fusion with the base metal.

(b) For flare-bevel and flare-groove welds, the etched cross-sections shall have complete fusion and the designated effective throat. The welds shall have no cracks in either the weld metal or heat-affected zone.

2.5.6 Penetrant Examination (Appendix F)

(1) For weld cladding, the entire clad surface of the test weldment shall be penetrant examined. To pass the test, there shall be:

(a) No linear indication with major dimensions greater than 1/16 in. (1.6 mm). A linear indication is one in which the length is more than three times the width.

(b) No more than three rounded indications with dimensions greater than 1/16 in. (1.6 mm) in a line separated by at least 1/16 in. (1.6 mm).

(2) For hardfacing welds, the hardfaced area of the test weldment shall be penetrant examined and shall meet the acceptance standards required by the referencing document.

2.5.7 Shear Test. Unless otherwise specified by the referencing document, to pass the test, the shear strength of each specimen shall be not less than 60 percent of the specified minimum base or filler metal tensile strength (whichever is lower).

2.5.8 Hardness Tests. To pass the hardness examination, a minimum of three readings shall be taken at the minimum weld metal thickness to be qualified on the WPS and shall not be less than the hardness specified in the referencing document.

2.5.9 Torque Tests

(1) Stud Welds - To pass the test(s), each of five welded studs shall support the required torque shown in Table 2.5.9 before failure occurs.

Table 2.5.9
Required torque for testing threaded unlubricated steel studs

Nominal diameter of studs		Threads per inch and series designated	Testing torque	
in.	mm		ft-lb	J
1/4	6.4	28 UNF	5.0	6.8
1/4		20 UNC	4.2	5.7
5/16	7.9	24 UNF	9.5	12.9
5/16		18 UNC	8.6	11.7
3/8	9.5	24 UNF	17.0	23.0
3/8		16 UNC	15.0	20.3
7/16	11.1	20 UNF	27.0	36.6
7/16		14 UNC	24.0	32.5
1/2	12.7	20 UNF	42.0	57.0
1/2		13 UNC	37.0	50.2
9/16	14.3	18 UNF	60.0	81.4
9/16		12 UNC	54.0	73.2
5/8	15.9	18 UNF	84.0	114.0
5/8		11 UNC	74.0	100.0
3/4	19.0	16 UNF	147.0	200.0
3/4		10 UNC	132.0	180.0
7/8	22.2	14 UNF	234.0	320.0
7/8		9 UNC	212.0	285.0
1	25.4	12 UNF	348.0	470.0
1		8 UNC	318.0	430.0

2.5.10 Fracture Toughness Tests

(1) The type of test, location, number of specimens, and acceptance criteria for fracture toughness tests shall be in accordance with the referencing document requiring such tests.

(2) Test procedures and apparatus for Charpy V-notch impact tests shall conform to the requirements of ASTM A370.

(3) Test procedures and apparatus for drop weight tests shall conform to the requirements of ASTM E208.

2.5.11 Special Tests Weldments. Acceptance criteria shall be in accordance with the referencing document.

2.6 Welding Variables

The following matrix indicates the welding variables to be included in a WPS and documented on a PQR for each welding process. A WPS or PQR may be presented in any format, written or tabular, provided the data required in this matrix are included (see 2.1.3). The PQR shall list the actual variables used within the limits of a narrow range. The WPS may list these same variables within the full range permitted for qualification variables and practical limits determined by the employer for other than qualification welding variables.

Welding Process*

Welding Variable	OFW	SMAW	GTAW	SAW	GMAW	FCAW	PAW	ESW	EGW	EBW	SW
2.6.1 Joint Design (reference 2.7.1)											
Joint type and dimensions	X	X	X	X	X	X	X	X	X	X	
Treatment of backside/method of gouging		X	X	X	X	X	X	X	X	X	
Backing and backing material	X	X	X	X	X	X	X	X	X	X	
Size, shape, ferrule/flux type											X
2.6.2 Base Metal (reference 2.7.2)											
Material number/subgroup/thickness	X	X	X	X	X	X	X	X	X	X	X
Surfacing metal and thickness		X	X	X	X	X	X	X	X	X	X
Product form (diameter, if pipe)	X	X	X	X	X	X	X	X	X	X	X
2.6.3 Filler Metal (reference 2.7.3)											
Classification, specification, F Number, and A Number	X	X	X	X	X	X	X	X	X	X	
Nominal composition, if not classified	X	X	X	X	X	X	X	X	X	X	
Filler metal/electrode diameter	X	X	X	X	X	X	X	X	X	X	
Flux classification	X			X		X					
Supplemental filler metal				X	X	X	X	X	X	X	X
Consumable insert and type			X				X				
Consumable guide								X			
Supplemental deoxidant											X
2.6.4 Position (reference 2.7.4)											
Welding position and vertical progression	X	X	X	X	X	X	X	X	X	X	X
2.6.5 Heat Treatment (reference 2.7.5 and 2.7.6)											
Preheat minimum		X	X	X	X	X	X			X	X
Interpass maximum, and postweld maintenance		X	X	X	X	X	X			X	
PWHT temperature and time at temperature	X	X	X	X	X	X	X	X	X	X	X
2.6.6 Gas (reference 2.7.7)											
Shielding gas and flow rate			X		X	X	X		X		
Torch root shielding gas and flow rate			X				X				
Fuel gas and flame type (oxidizing, neutral, or reducing)	X										
Environmental shielding and pressure										X	
2.6.7 Electrical (reference 2.7.8)											
Current (or wire feed speed), current type, and polarity		X	X	X	X	X	X	X	X	X	X
Voltage		X	X	X	X	X	X	X	X	X	
Beam focus current, pulse freq., filament, type, shape, and size										X	

*OFW = Oxyfuel Gas Welding
SMAW = Shielded Metal Arc Welding
GTAW = Gas Tungsten Arc Welding
SAW = Submerged Arc Welding
GMAW = Gas Metal Arc Welding
FCAW = Flux Cored Arc Welding
PAW = Plasma Arc Welding
ESW = Electroslag Welding
EGW = Electrogas Welding
EBW = Electron Beam Welding
SW   = Stud Welding

Welding Process (cont'd)*

Welding Variable	OFW	SMAW	GTAW	SAW	GMAW	FCAW	PAW	ESW	EGW	SW
Type and size of tungsten electrode			X				X			
Transfer mode					X	X				
Pulsing			X		X	X	X			

2.6.8 Technique (reference 2.7.9)

Welding Variable	OFW	SMAW	GTAW	SAW	GMAW	FCAW	PAW	ESW	EGW	SW
Process and whether manual, semi-automatic, or automatic/machine	X	X	X	X	X	X	X	X	X	X
Single or multi electrode and spacing		X	X	X	X	X	X	X		
Single or multi pass (per side), single or double weld	X	X	X	X	X	X	X	X	X	X
Contact tip to work distance				X	X	X		X	X	
Oscillation variables (automatic/machine)			X	X	X	X	X	X	X	
Peening	X	X	X	X	X	X	X	X	X	
Conventional or keyhole technique							X			
Cleaning	X	X	X	X	X	X	X	X	X	X
Stud gun model and lift										X
Vacuum or non-vacuum, and gun-to-work distance and gun angle									X	
Backing shoe type								X	X	
Stringer or weave bead		X	X	X	X	X	X			
Travel speed			X	X	X	X	X	X	X	X

2.7 Qualification Variables

This paragraph lists the qualification variables for each welding process covered by this document.

Qualification Variable - Those variables which, if changed beyond the limitations given below, require requalification of the procedure as designated by the following symbols:

Q - Qualification Variable for all applications

T - Qualification Variable for fracture toughness applications

C - Qualification Variable for Weld Cladding applications

H - Qualification Variable for Hardfacing applications

Welding Process*

Qualification Variable	OFW	SMAW	GTAW	SAW	GMAW	FCAW	PAW	ESW	EGW	SW

2.7.1 Joint Design

(1) A change in the joint design type from one of the listed types to another type.

    (a) Square groove

    (b) Single bevel or J-groove

    (c) U or V-groove, double U- or V-groove

(2) If backing is used, a change in M Number of the backing.

Qualification Variable	OFW	SMAW	GTAW	SAW	GMAW	FCAW	PAW	ESW	EGW	SW
2.7.1 (1)(c)	Q	Q	Q	Q	Q	Q	Q	Q	Q	Q
2.7.1 (2)	Q	Q	Q	Q	Q	Q	Q	Q	Q	Q

Qualification Variable

(3) The addition of thermal back gouging on M-11 materials.

(4) A change in root face thickness exceeding 20%.

2.7.2 Base Metal

(1) A change in the base metal thickness beyond the range qualified in 2.4.1.

(2) A change from one M Number to another M Number or to an unlisted base metal.

(3) A change in Group Number to another Group Number within the same M Number.

(4) For joints between base metals of different M Numbers, requalification is required even though the base metals have been separately qualified except as detailed in (5) and (6) below.

(5) For M Numbers 1, 3, 4, and 5 (of 3% maximum nominal chromium content), a procedure qualification test (excluding PAW keyhole techniques) with one M Number shall also qualify for that metal welded to each of the lower M Number metals, but not higher.

(6) Where fracture toughness is required, qualification shall be made using base metal(s) from the same M Number(s) used in production; if however, procedure qualification tests have been made for each of the two base metals welded to itself, using the same procedure including the same qualified variables, it shall only be necessary to prepare a WPS showing the combination of base metals.

(7) A change from M-9A to M-9B, but not vice versa. A change from one group of M-10 to any other group of M-10.

(8) A change from one Group Number of M-11 to any other group.

(9) A decrease in the thickness or change in nominal specified chemical analysis of weld metal beyond that qualified.

(10) A change in the nominal size or shape of the stud at the section to be welded.

Welding Process*

Variable	OFW	SMAW	GTAW	SAW	GMAW	FCAW	PAW	ESW	EGW	EBW	SW
(3) thermal back gouging	Q	Q	Q	Q	Q	Q	Q	Q	Q	Q	
(4) root face	Q									Q	
2.7.2 (1) base metal thickness	Q	Q	Q	Q	Q	Q	Q	Q	Q	Q	
(2) M Number	Q	Q	Q	Q	Q	Q	Q	Q	Q	Q	Q
(3) Group Number		T	T	T	T	T	T	T	T		
(4) joints different M	Q	Q	Q	Q	Q	Q	Q	Q	Q	Q	
(5) M Numbers 1,3,4,5		Q	Q	Q	Q	Q					
(6) fracture toughness		T	T	T	T	T	T	T	T		
(7) M-9A to M-9B	Q	Q	Q	Q	Q	Q	Q	Q	Q	Q	
(8) Group of M-11	Q	Q	Q	Q	Q	Q	Q				
(9) decrease thickness	Q	Q	Q	Q	Q	Q				Q	
(10) nominal size of stud											Q

---

*OFW	=	Oxyfuel Gas Welding	PAW	=	Plasma Arc Welding
SMAW	=	Shielded Metal Arc Welding	ESW	=	Electroslag Welding
GTAW	=	Gas Tungsten Arc Welding	EGW	=	Electrogas Welding
SAW	=	Submerged Arc Welding	EBW	=	Electron Beam Welding
GMAW	=	Gas Metal Arc Welding	SW	=	Stud Welding
FCAW	=	Flux Cord Arc Welding			

Qualification Variable (cont'd.)

2.7.3 Filler metals

(1) A change from one F Number to any other F Number or to any filler metal not listed in Appendix B.

(2) For ferrous filler metals, a change from one A Number to any other A Number, or to a filler metal analysis not listed in Appendix B (The PQR and WPS shall state the nominal chemical composition and manufacturers designation identifying filler metals which do not fall into an A Number group). Qualification with A-1 shall qualify for A-2 and vice versa.

(3) For surfacing, a change in the nominal composition of the weld metal A Number, or alloy type (each layer is considered independent of other layers).

(4) A change in nominal filler metal strength exceeding 10,000 psi, or a change in filler metal specified UTS as indicated by the electrode classification number, or as specified in the applicable filler metal specification to a strength lower than the minimum specified ultimate strength of the base metal.

(5) Where the alloy content of the weld metal is largely dependent upon the composition of the flux, any change in the welding procedure which would result in the important alloying elements in the weld metal being outside of the specified range of chemical composition given in the WPS.

(6) A change in the nominal size/shape of filler metal/electrode in the first layer.

(7) The addition or deletion of supplemental filler metal (powder or wire), or a change of 10% in the amount.

(8) A change from single to multiple supplementary filler metal or vice versa.

(9) A change in filler metal from bare (solid) to flux cored, to flux covered, or vice versa.

(10) The addition or deletion of consumable inserts.

(11) A change from consumable guide to nonconsumable guide, and vice versa.

(12) Addition or deletion, or a change in nominal amount or composition of supplementary metal (in addition to filler metal) beyond that qualified.

(13) A change from wire to strip electrodes and vice versa.

(14) A change from one flux classification listed in an AWS specification to any other flux classification, or to any unlisted flux. Variation of molybdenum content of the weld metal ± 0.5% does not require requalification.

Welding Process (cont'd)*									
OFW	SMAW	GTAW	SAW	GMAW	FCAW	PAW	ESW	EGW	SW
Q	Q	Q	Q	Q	Q	Q	Q	Q	
Q	Q	Q	Q	Q	Q	Q	Q	Q	Q
CH	CH	CH	CH	CH	CH	CH			
Q	Q	Q	Q	Q	Q	Q	Q	Q	Q
				Q			Q		
	CH			CH	CH	CH			
Q	T	T	T	T	T	T	T	T	Q
CH			CH	CH	CH	CH	CH		
				Q			Q		
			Q			Q			
							Q		
				Q			Q		Q
							Q		
			Q	Q	Q				

Qualification Variable (cont'd.)

2.7.4 Position

(1) A change from any position to the vertical position, uphill progression. Vertical uphill qualifies for all positions.

(2) The addition of welding positions other than that qualified except that positions other than flat also qualify for flat.

2.7.5 Preheat/Interpass Temperature

(1) A decrease in temperature of more than 100F from that qualified and recorded on the PQR.

(2) An increase in temperature of more than 100F from that qualified and recorded on the PQR.

2.7.6 Postweld Heat Treatment

(1) A separate PQR is required for each of the following postweld heat treatment conditions specified in the WPS:

    (a) No postweld heat treatment.

    (b) Postweld heat treatment below, within, or above (e.g., normalized) the critical temperature range.

    (c) Postweld heat treatment above the critical temperature range followed by postweld heat treatment below the critical range (e.g., normalized and tempered).

    (d) For M-8 metal, the addition or deletion of a solution or stabilizing heat treatment.

(2) For M-11 metals and for metals where fracture toughness is a requirement, a change in the specified postweld heat treatment temperature [see 2.7.6(1)] and time range requires a procedure qualification. The qualification test weldment shall be subjected to heat treatment essentially equivalent to that used on the production weldment, including at least 80% of the aggregate times at temperature(s).

2.7.7 Gas Shielding

(1) The omission of root shielding gas.

(2) The omission of shielding gas.

(3) A change in shielding gas from a single gas to any other single gas or to a mixture of gases, or a change in specified nominal percentage composition of gas mixture.

(4) Any change of environmental shielding such as from vacuum to an inert gas, or vice versa. An increase in the absolute pressure of the vacuum welding environment beyond that qualified.

(5) A change in shielding as a result of ferrule or flux type.

Welding Process (cont'd)*

O	S	G	S	G	F	P	E	E	S	
F	M	T	A	M	C	A	S	G	B	W
W	A	A	W	A	A	W	W	W	W	
	W	W		W	W					
T	T			T	T	T				
CH	CH	CH	CH	CH	CH	CH				
	Q	Q	Q	Q	Q	Q			Q	Q
T	T	T	T	T	T	T	T	T		
TH	T	T	T	T	T	T	T	T		
TH	T	T	T	T	T	T	T	T		
TH	T	T	T	T	T	T	T	T		
TH	T	T	T	T	T	T	T	T		
TH	T	T	T	T	T	T	T	T		
				Q		Q	Q	Q		
				Q		Q	Q	Q		Q
H				Q		Q	Q	Q		Q
										Q
										Q

Qualification Variable (cont'd.)

(6) For M Numbers 51, 52, 61, and 10-I metals, a change in the nominal composition or a decrease of 15% or more from the flow rate of root shielding gas.

2.7.8 Electrical Characteristics

(1) Except when the WPS is qualified with a grain-refining austenitizing heat treatment after welding, an increase in heat input or volume of weld metal bead per unit length of weld over that qualified. The increase may be measured by either of the following:

(a) Heat input (J/in.) =

$$\frac{\text{Voltage} \times \text{Amperage} \times 60}{\text{Travel Speed (in./min)}}$$

(b) Volume of weld metal - An increase in bead size, or a decrease in length of weld bead per unit length of electrode, or a decrease in travel speed.

(2) A change of plus or minus 15% from the amperage or voltage ranges qualified, or a change in the beam pulsing frequency duration from that qualified, or any change in filament type, size, or shape.

(3) A parameter change from that qualified:

(a) Exceeding ± 2% in the voltage or welding speed.

(b) Exceeding ± 5% in the beam current, beam focus current, or gun-to-work distance.

(c) Exceeding ± 20% in oscillation length or width.

(d) In the beam pulsing frequency duration.

(4) A change in the type of power source, or a change in the arc timing of more than ± 1/10 second. A change in amperage of more than ± 10%.

(5) A change in the mode of metal transfer; globular, spray, or short-circuiting.

(6) Addition or deletion of pulsing current to dc power source.

2.7.9 Other

(1) A change in welding process.

(2) A change from single electrode to multiple electrode, or vice versa. (Multiple electrode is defined as two or more electrodes acting in the same weld puddle.)

(3) A change from multi-pass per side to single pass per side.

Welding Process (cont'd)*

Qualification Variable	O F W	S M A W	G T A W	S A W	G M A W	F C A W	P A W	E S W	E G W	S W
2.7.7 (6)			Q		Q		Q			
2.7.8 (1)(b)		T	T	T	T	T	T	T	T	
2.7.8 (2)									Q	
2.7.8 (3)(d)									Q	
2.7.8 (4)										Q
2.7.8 (5)					Q					
2.7.8 (6)			Q		Q	Q				
2.7.9 (1)	Q	Q	Q	Q	Q	Q	Q	Q	Q	Q
2.7.9 (2)			CH	T	T	T	T	T	Q	Q
2.7.9 (3)		T	T	T	T	T				

Welding Process (cont'd)*

```
O  S  G  S  G  F  P  E  E  S
F  M  T  A  M  C  A  S  G  B  W
W  A  A  W  A  A  W  W  W  W
   W  W     W  W
```

OFW	SMAW	GTAW	SAW	GMAW	FCAW	PAW	ESW	EGW	EBW	SW
						T				
									Q	
										Q
										Q
									Q	
							Q	Q		
							Q	Q		
		T	T	T	T	T				
		CH	CH	CH	CH	CH	Q	Q		
	CH	CH	CH	CH	CH					
	CH	CH	CH	CH	CH	CH				
Q										

Qualification Variable (cont'd.)

(4) A change from the conventional technique to the keyhole technique of welding or vice versa, or the inclusion of both techniques even though each has been individually qualified.

(5) A change from vacuum to nonvacuum welding in M Number 50 or 60 series metals, but not vice versa.

(6) A change in the following stud welding conditions:

    (a) A change of stud gun model.

    (b) A change exceeding 1/32 in. in the nominal lift.

(7) A change from that qualified, in gun-to-work distance of more than ± 5%, or change in oscillation length or width if more than ± 20%, or a change in the gun angle (axis of the beam), or the addition of a wash pass, or a change exceeding 2% of the qualified travel speed.

(8) The addition or deletion of backing shoes or nonfusing metal retainers.

(9) A change in design and/or material of backing shoes, either fixed or movable, from nonfusing solid to water cooled or vice versa.

(10) A change exceeding 20% in the oscillation variables.

(11) A change in travel speed range for machine or automatic welding of 10% or more.

(12) A change from stringer bead to weave bead for manual or semi-automatic.

(13) A change in the flame type (oxidizing, neutral, or reducing flame, or vice versa).

2.8 Standard Test Weldments - Location and Number of Test Specimens

The standard test weldments with location of test specimens are shown in Figs. 2.8.1, 2.8.2, 2.8.3, 2.8.4A, 2.8.4B, 2.8.5, and 2.8.6. The type of tests required for procedure qualification are given in Table 2.2.2.

2.8.1 Standard Pipe Weldment. The test weldment shall be composed of two pipe sections, a minimum of 5 in. long, joined together by welding to make a test weldment of 10 in. (254 mm) min. length. Where only one test weldment is prepared, the diameter shall be sufficient to accommodate all required specimens. The thickness shall be based upon the requirements of Table 2.4.1A.

2.8.1.1 Required Specimens. Required specimens shall be removed from approximate locations as shown in Fig. 2.8.1.

2 Reduced Section Tensile*
2 Face Bends**
2 Root Bends**

Side bend specimens may be substituted for root and face bend specimens for metal thicknesses from 3/8 in. (9.5 mm) to 3/4 in. (19.0 mm) inclusive. For metal over 3/4 in. (19.0 mm) thick, side bend specimens shall be used. When toughness testing is a requirement, it shall be applied with respect to each welding process and

---

  *The weld metal of each welding process and of each welding procedure shall be included in the tension test specimen.

  **The weld metal of each welding process and of each welding procedure shall be included on the tension side of the bend when the face and root bend samples are used.

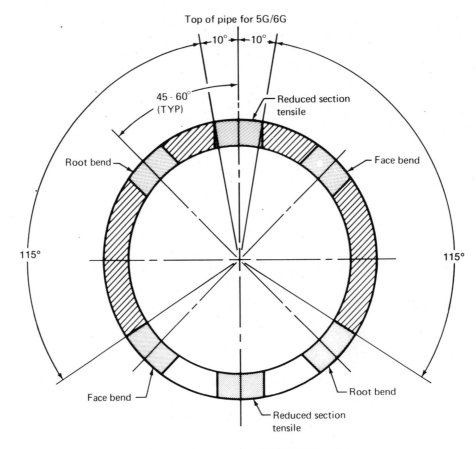

Notes:

1. Toughness specimens, when required, shall be removed from the ▨▨▨▨ sections.
2. Side bend specimens may be substituted for face and root bends for metal thickness from 3/8 in. to 3/4 in., inclusive. For metal thicknesses over 3/4 in., side bends shall be used.

Fig. 2.8.1 — Approximate location of test specimens for pipe

welding procedure.

2.8.2 Standard Plate Weldment. Test specimens are normally transverse. Longitudinal specimens may be used in lieu of transverse bend specimens for material combinations which differ markedly in bending properties between the two base metals or the weld metal and base metal as shown in Fig. 2.8.2.

2.8.2.1 Required Specimens:
2 Tension Tests (Reduced Section)
2 Root Bends
2 Face Bends

Side bend specimens may be substituted for root and face bend specimens for metal thicknesses from 3/8 in. (9.5 mm) to 3/4 in. (19.0 mm) inclusive. For metal over 3/4 in. (19.0 mm) thick, side bend specimens shall be used.

2.8.3 Standard Fillet Weld Test Weldment. The standard fillet test weldment is shown in Fig. 2.8.3. Length should be sufficient for required number of test specimens, which may be of any convenient width not less than 1 in. (25.4 mm).

2.8.3.1 Required Specimens:
4 Transverse Shear Tests

2.8.4 Standard Reinforcing Bar Test Weldments. Three test weldments shall be made using one of the applicable joint designs shown in Figs. 2.8.4A or B, and shall be used for full section tensile tests and macroetch test.

2.8.4.1 Required Specimens:
1 Macroetch
2 Tension Tests

2.8.5 Standard Cladding Test Weldment. The test weldment is shown in Fig. 2.8.5. The base metal shall be within the thickness limitation specified in Table 2.4.1D, Note 9.

2.8.5.1 Test Specimens:
Penetrant Examination - On the minimum weld clad thickness qualified
Side Bends - (4 Transverse or 2 Transverse and 2 Longitudinal)

2.8.6 Standard Hardfacing Test Weldments. The test weldment is shown in Fig. 2.8.6. The thickness shall be based on the requirements of Table 2.4.1D.

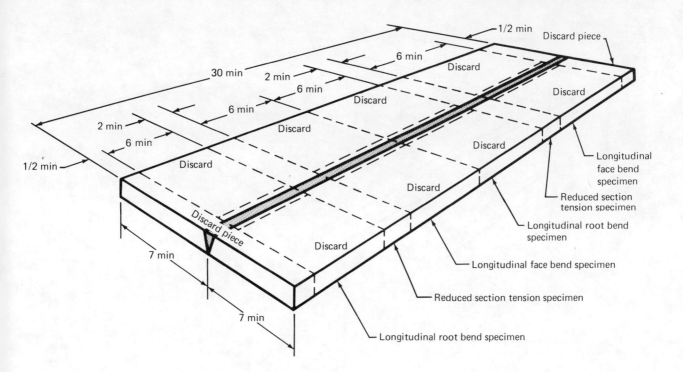

Longitudinal bend specimens

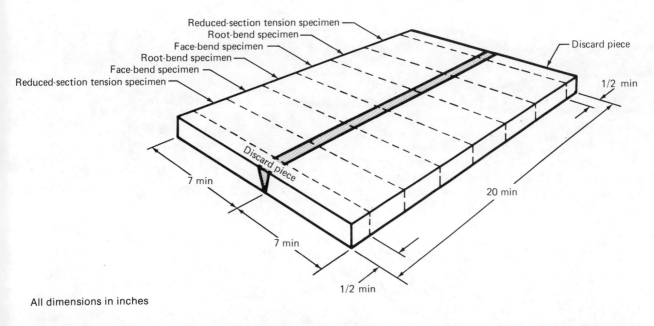

All dimensions in inches

Transverse bend specimens

Notes:

1. T = Test plate thickness per Table 2.4.1A.
2. Side bend specimens may be substituted for root and face bend specimens for metal thicknesses from 3/8 in. to 3/4 in., inclusive. For metal over 3/4 in. thick, side bend specimens shall be used.

Fig. 2.8.2 — Location of test specimens in plate groove test weldments

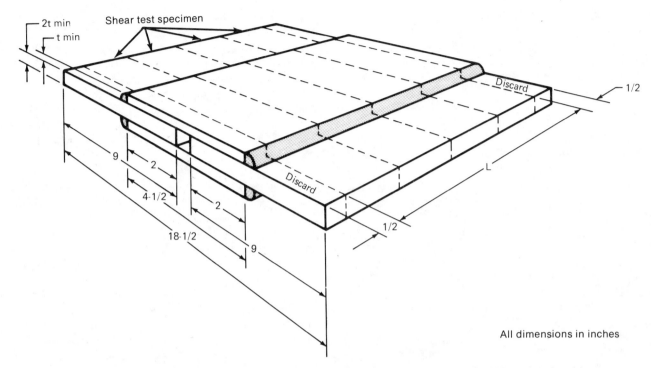

All dimensions in inches

L = Length should be sufficient for the required number of specimens, which may be of any convenient width not less than 1 in.

t = Specified fillet weld size plus 1/8

Fig. 2.8.3 — Standard fillet weld test weldment

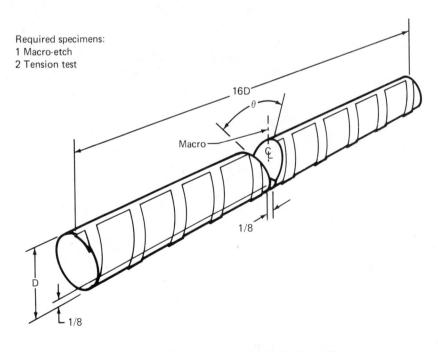

Required specimens:
1 Macro-etch
2 Tension test

(1) For Bars No. 9 or larger, use single-V or bevel-groove welds (0=45 to 60)

(2) For Bars No. 8 or smaller, use single-V with split-pipe backing (0=60)

Fig. 2.8.4A — Butt joint in reinforcing bar

2.8.6.1 Test Specimens:

Penetrant Examination - 1 in. x 4 in. (25.4 mm x 101 mm) surface/min. qualified thickness.

Macroetch Examination - Two transverse sections, 3 Hardness Readings on the minimum thickness qualified.

Chemical Analysis (Appendix A, Fig. A3B.)

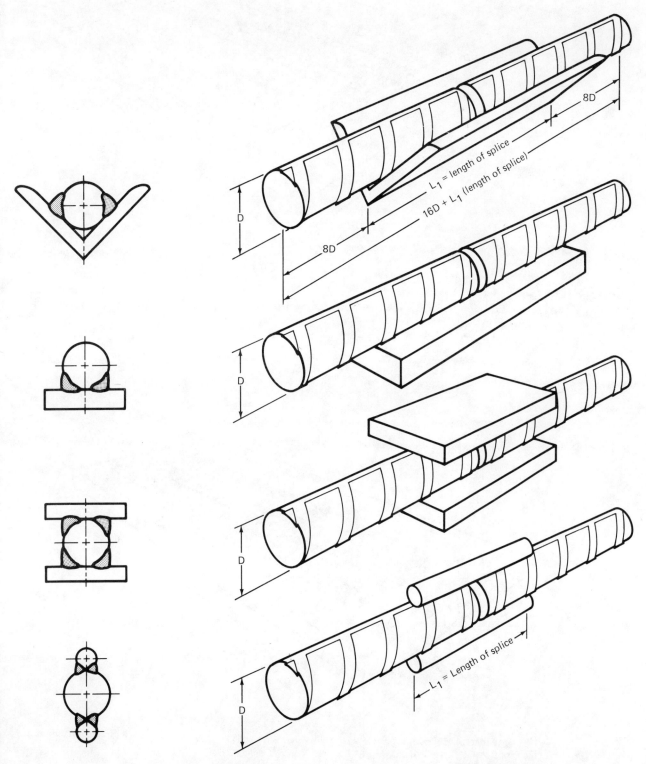

Fig. 2.8.4B — Spliced butt joint (flare groove) in reinforcing bar

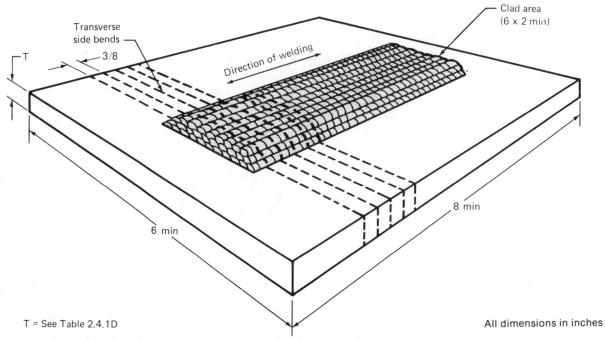

T = See Table 2.4.1D

Note: The number of passes shown is illustrative only.

All dimensions in inches

Fig. 2.8.5 — Cladding test weldment

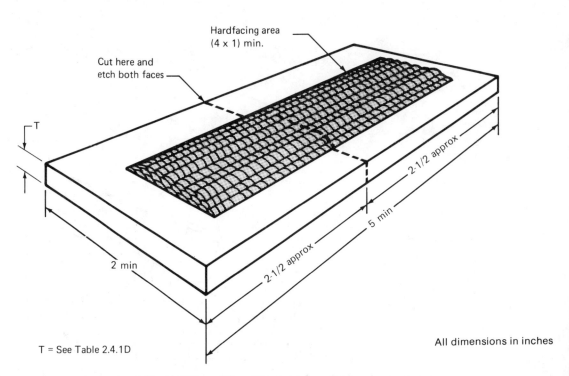

T = See Table 2.4.1D

All dimensions in inches

Fig. 2.8.6 — Hardfacing specimen

3. Performance Qualification

3.1 General

3.1.1 Qualification under this Section requires completion of the specified test weldment and acceptance by the Qualifier of the test weldment and test results.

3.1.2 Acceptance of performance qualification weldments shall be allowed by either of two methods (see Fig. 3.1.2):

    (1) Qualification by workmanship test (See 3.2)

    (2) Qualification by standard test (See 3.3)

3.1.3 Qualification as a welder does not qualify the individual as a welding operator or vice versa.

3.1.4 Performance qualification by standard test shall qualify the individual to perform welding where qualification by either standard test or workmanship test is specified. Performance qualification by workmanship test shall qualify for production work only when qualification by visual examination is permitted by the referencing document.

3.1.5 Qualification for weld cladding qualifies only for weld cladding. Qualification for hardfacing shall qualify only for hardfacing.

3.1.6 Qualification to 3.2 and 3.3 is permitted on production weldments provided such qualification is permitted by the referencing document.

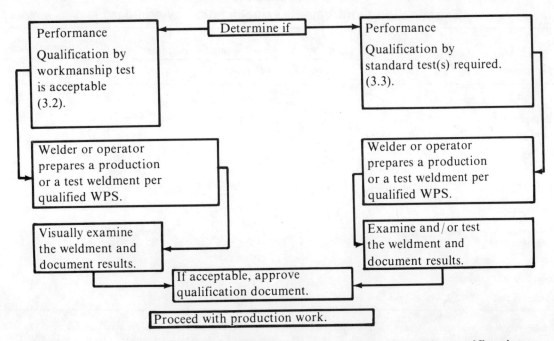

Fig. 3.1.2 — Performance qualification steps required by this specification

3.1.7 A welder or welding operator who completes an acceptable procedure or performance qualification test weldment shall be qualified for production work within the limits of variables specified in 3.6.

3.1.8 The welder or welding operator undertaking performance qualification tests shall be under the full supervision and control of the qualifier during the welding of test weldments. Acceptance or rejection, and documentation of test results is the responsibility of the qualifier. Performance qualification documentation shall be dated and signed by the qualifier.

3.1.9 The performance qualification shall remain in effect indefinitely unless the welder or welding operator is not engaged in the process qualified for a period exceeding 12 months or there is some specific reason to question a welder's, or welding operator's, ability. In these instances, re-qualification may be accomplished by making and testing one test weldment using the same process.

3.1.10 A welder or welding operator who fails the performance test required by this Article may be retested at the option of the qualifier, as specified below:

(1) If an immediate retest is permitted by the qualifier, it shall consist of 2 test weldments for each position failed. Both test weldments must be acceptable.

(2) Provided there is evidence satisfactory to the qualifier that the welder or welding operator has had further training or practice, a retest consisting of all the original test requirements may be permitted.

3.2 Qualification by Workmanship Test

3.2.1 Qualification by workmanship test is permitted when allowed by the referencing document.

3.2.2 Qualification by workmanship test requires completion of a workmanship weldment in accordance with a qualified WPS, representing typical production configurations and conditions, which shall be accepted or rejected primarily by visual examination and other examinations (i.e. macro-examination etc.) as may be deemed appropriate for the application. Figure 3.7.6 illustrates typical workmanship test weldments.

3.2.3 Tables 3.4.6A and 3.4.6B define the positions qualified based upon the position tested. Weldment orientation in other than the standard (1G, 2G, etc.) positions is permitted. Such qualifications are valid only for the position tested. Angular deviation in weld axis inclination and face rotation shall be in accordance with Figs. A1.1A and A1.1B.

3.2.4 The thickness (and diameter where applicable) of the test weldment shall be based upon the weld

Table 3.3.1
Examination requirements - qualification by standard test

Examination/testing	Pipe or plate		Reinforcing bar butt joint		Surfacing	
	Groove	Fillet	Vee groove	Flare groove	Clad	Hardface
Visual examination	Yes	Yes	Yes	Yes	Yes	Yes
Radiography examination*	Yes, or bend	—	Yes	—	—	—
Bend test	Yes, or radiograph	—	—	—	Yes	—
Break test	—	Yes	—	—	—	—
Macro-examination	—	Yes	Yes, or radiograph	Yes	—	—
Tension test	—	—	Yes, or radiograph	—	—	—
Penetrant examination	—	—	—	—	Yes	Yes
Hardness test	—	—	—	—	—	Yes

*Radiographic examination shall be performed in accordance with Appendix D of this Specification, and is acceptable as an alternative to bend tests for the following processes: SMAW, GTAW, GMAW (except for the short-circuiting transfer mode), FACW, PAW, and SAW.

thickness to be welded in production. Tables 3.4.5 A, B, C, D, and E show the ranges qualified by a given test weldment thickness or diameter, or both.

3.2.5 The test weldment shall be visually examined and shall satisfy the requirements of 3.5.1, except that where partial penetration welds are specified, the requirements of 3.5.1.(2) do not apply.

3.3 Qualification by Standard Test

3.3.1 Performance qualification by standard test shall be in accordance with Table 3.3.1.

3.3.2 Qualification by standard test requires completion of a standard test weldment (see 3.7) in accordance with a qualified WPS.

3.3.3 Tables 3.4.6A and B define the positions qualified based upon position tested. Weldment orientation other than the standard (1G, 2G, etc.) positions is permitted, but such tests qualify only for the position tested. Angular deviation in weld axis inclination and face rotation shall be in accordance with Figs. A1.1A and A1.1B.

3.3.4 The thickness (and diameter where applicable) of the test weldment shall be based upon the weld thicknesses and diameters to be welded in production. Tables 3.4.5A, B, C, D, and E show the ranges quali-

fied by a given test weldment thickness or diameter, or both.

3.3.5 The test weldment shall satisfy the applicable acceptance criteria of 3.5.

3.4 Test Weldments

3.4.1 Standard test weldments shall meet the dimensional requirements for test weldments (3.7) and shall be welded in one or more of the standard test positions (Appendix A).

3.4.2 Test weldments may also be workmanship samples (See 3.7.6) if qualification is performed under the requirements of 3.3.

3.4.3 Multi-process qualification on the same test weldment is permitted. The qualified thickness range and pipe diameter size for each welding process shall be in accordance with tables shown in 3.4.5. The weld metal thickness for each welding process shall be documented along with all other variables on the Performance Qualification Record (see 3.6).

3.4.4 Qualification shall be performed using the metals (Appendices B and C, F Number and M Number respectively) to be used in production, which shall qualify only for metals under the same M Number or F Number, except that some metals shall qualify for other metals as specified in Table 3.4.4.

Table 3.4.4
Allowable test weldment material groups

Metal used for performance test weldment		Qualifies for	
Base metal M Number(s) (Note 2)	Filler metal F Number (Note 2)	Base metal M Number	Filler metal F Number (Note 1)
1 through 11	—	All, except 20, 30, 50 or 60 series metals	—
20 Series	—	Any 20 Series metal	—
30 Series	—	Any 30 Series metal	—
	1 through 5 and 41 through 43	—	The F Number used in test weldment and any lower F Number 1 through 5, and 41 through 43
	21, 22, 23	—	Qualify for themselves and each other

Notes:
1. Welding operator performance tests made using any filler metal for a process shall be qualified for any other filler metal.
2. Performance tests using a base metal not listed in Appendix C shall qualify only for that base metal. Tests using a filler metal not listed in Appendix B shall qualify for using similar listed filler metals of the nominal chemical composition for the process used in the test.

Table 3.4.5A
Pipe groove test weldments to qualify for
production welding of pipe and plate

Test weldment	Qualifies for pipe or plate					
	Outside diameter		Thickness			
			Groove welds			
Outside diameter, in. (mm)	Groove welds	Fillet welds	Min	Max	Fillet welds	
Less than 1 (25.4)	Size welded and over	All	1/2t	2t		
1 (25.4) up to 2-7/8 (73) incl.	1 in. (25.4 mm) and over	All	1/2t	2t	All	
2-7/8 (73) up to 6-5/8 (168) incl.	2-7/8 (73 mm) and over	All	1/2t	2t		
Over 6-5/8 (168)	4-1/2 (114 mm) and over	All	1/2t	No max		

t = thickness of weld metal of test weldments (not including reinforcement).

Table 3.4.5B
Plate groove test weldments to qualify for
production welding of plate and sheet

Test weldment	Qualifies for				
	Groove welds Thickness, in. (mm)		Fillet welds Leg size, in. (mm)		
Plate thickness (T), in. (mm)	Min	Max	Min	Max	Pipe
To 3/8 incl. (9.5)	1/2t	2t	1/2t	No max	Not qualified
Over 3/8 (9.5)	3/16 (4.8)	Max to be welded	3/8 (4.8)	No max	

t = thickness of weld metal of test weldment (not including reinforcement).

Note: Qualification on plate shall also qualify for pipe over 24 in., within the range of thickness allowed in Table 3.4.5A.

3.4.5 Tables 3.4.5A, B, C, D, and E define the range of thickness and pipe diameters qualified by a test weldment of any given size and thickness.

3.4.6 The positions for which a welder or welding operator becomes qualified when successfully completing a test weldment in one or more of the positions illustrated in Appendix A are shown in Tables 3.4.6A and B.

3.5 Examination and Acceptance. The minimum examination requirements are outlined as follows:

3.5.1 Visual Examination. The test weldment may be examined visually by the qualifier at any time, and the test may be terminated at any stage if the welder or welding operator does not exhibit the necessary skills. Visual examination means examination without magnification other than corrective lens. Acceptance criteria for visual examination are:

(1) There shall be no cracks or incomplete fusion.

(2) There shall be no incomplete joint penetration in groove welds, except for partial joint penetration groove welds (see 3.2.5).

(3) The appearance of the weld shall satisfy the qualifier that the welder is skilled in using the process and procedure specified for the test.

Table 3.4.5C
Fillet test weldments qualification limits — diameter and weld thickness

	Qualifies for			
Test weldment	Outside diameter (No maximum) in. (mm)		Weld thickness range qualified fillet leg size, in. (mm) plate or pipe	
Weldment, in. (mm)	Fillet (Note 1)	Groove	Min	Max
Pipe: (Outside diameter) Less than 1 (25.4)	Size welded and over	Not qualified	No min	No max
1 (25.4) through 2-7/8 (73)	1 (25.4) and over			
Over 2-7/8 (73)	2-7/8 (73) and over			
Plate: Thickness (T) 3/16 (4.8) and less (Note 1)	2-7/8 (73) and over	Not Qualified	1/2T	2T
Over 3/16 (4.8)			3/32 (2.4)	No max

T = Thickness of test plate.

Note 1. Fillet weld qualifications on plate are valid for pipe fillets or sizes 2-7/8 in. or over. For diameters less than 2-7/8 in., qualification shall be on a pipe fillet test weldment or a groove test weldment.

Table 3.4.5D
Rebar test qualification limitations on rebar size

	Qualifies for Bar Size					
Rebar test weldment	Groove Complete penetration		Splices Partial joint penetration		Fillet and lap splices	
	Min	Max	Min	Max	Min	Max
Direct butt joint	Size welded	Size welded	No min	No max	No min	No max
Thermit/pressure weld	No min	No max	Not qualified		Not qualified	
Spliced butt joint (Flare bevel groove)	Not qualified		Size welded	Size welded	Size welded	Size welded

Table 3.4.5E
Weld cladding and hardfacing thickness qualification limitations

	Qualifies for			
Test weldment base metal thickness, in. (mm)	Base metal thickness, in. (mm)		Weld metal thickness, in. (mm)	
	Min	Max	Min	Max
Less than 1 (25.4)	T	No max	Same as WPS minimum qualified	No max
1 and over (25.4)	1 (25.4)	No max		

Table 3.4.6A
Limitations on groove weld positions qualified
(also includes surfacing welds)

Test position grooves		Positions qualified* (Notes 1, 2, and 3)						
		Groove welds		Reinforcing bar		Fillet welds		
		Pipe	Plate	Butt	Flare	Pipe	Plate	Rebar
1G	Pipe	F	F	—	F	F	F	F
	Plate	F	F	—	F	F	F	F
	Rebar Butt	—	—	F	—	—	—	F
	Flare	—	—	—	F	—	—	F
2G	Pipe	F,H	F,H	—	F,H	F,H	F,H	F,H
	Plate	F,H	F,H	—	F,H	F,H	F,H	F,H
	Rebar Butt	—	—	F,H	—	—	—	F,H
	Flare	—	—	—	F,H	—	—	F,H
3G	Plate	—	F,V	—	F,V	F,H	F,V,H	F,V,H
	Rebar Butt	—	—	F,V,H	—	—	—	F,V,H
	Flare	—	—	—	F,V,H	—	—	F,V,H
4G	Plate	—	F,O	—	F,O	F	F,H,O	F,H,O
	Rebar Butt	—	—	F,O	—	—	—	F,V,O
	Flare	—	—	—	F,O	—	—	F,H,O
5G	Pipe	F,V,O	F,V,O	—	F,V,O	F,V,O	F,V,O	F,V,O
6G	Pipe	All	All	—	All	All	All	All
6GR	Pipe	All	All	—	All	All	All	All
2G Plus 5G	Pipe	All	All	—	All	All	All	All
3G 4G	Plate	—	All	—	All	F,H	All	All

*F = Flat, H = Horizontal, V = Vertical, O = Overhead.

Notes:
1. Qualification on plate also qualifies for pipe over 24 in. (610 mm) diameter.
2. All position qualification for pipe over 2-7/8 in. (73 mm) O.D. may be performed on one, 6-5/8 in. (168 mm) diameter or greater, 2G/5G test weldment (Fig. 3.7.1C).
3. Qualification on double-welded plate also qualifies for double-welded pipe and vice versa.

Table 3.4.6B
Limitations on fillet weld
positions qualified

Test position fillets		Fillet positions qualified* (Note 1)		
		Pipe	Plate	Rebar
1F	Pipe	F	F	F
	Plate	F	F	F
2F/2FR	Pipe	F,H	F,H	F,H
	Plate	F,H	F,H	F,H
3F	Plate	—	F,V,H	—
4F	Pipe	F,H,O	F,H,O	F,H,O
	Plate	F,H,O	F,H,O	F,H,O
5F	Pipe	All	All	All
3F & 4F	Plate	All	All	All

*F = Flat, H = Horizontal, V = Vertical, O = Overhead.

Notes:
1. Qualification on plate also qualifies for pipe over 24 in. (610 mm) diameter.

(4) Undercut shall not exceed the lesser of 10% of the base metal thickness or 1/32 in. (0.8 mm).

(5) Where visual examination is the only criterion for acceptance, all weld passes are subject to visual examination.

3.5.2 Radiographic Examination - Groove Welds

3.5.2.1 Radiographic technique shall be based upon Appendix D of this Specification. Final interpretation of results is the responsibility of the qualifier. The test weldment, less any discard for plate, shall be fully examined.

3.5.2.2 For qualification on production work, acceptance shall be based on a minimum of 6 in. (150 mm) of production weld, and the acceptance criteria shall be per the document governing that production weld. However, if more than 6 in. (150 mm) appears on the radiographic film, the full length of the weld covered by the film shall be examined.

3.5.2.3 Face reinforcement may be removed at the option of the qualifier. Root reinforcement shall not be removed from single welded groove joints. Backing shall not be removed.

3.5.2.4 Acceptance shall be in accordance with the following requirements.

(1) Cracks, incomplete joint penetration, or incomplete fusion are not permitted.

(2) Linear discontinuities are those in which the length is more than three times the width. Permitted linear indications are shown in Table 3.5.2A.

Table 3.5.2A
Acceptable linear discontinuities

Weld thickness (t), in.	Max discontinuity, in.	Max aggregate length Aligned indications*
Up to 3/8, inclusive	1/8	
Over 3/8 to 2-1/4	1/3t	t in a length of 12t
Over 2-1/4	3/4	

Weld thickness (t), mm	Max discontinuity, mm	Max aggregate length Aligned indications*
Up to 9.5, inclusive	3.2	
Over 9.5 to 57	1/3t	t in a length of 12t
Over 57	19	

*Aligned indications are those where the distance between the successive indications in less than 6L, where L is the length of the longest indication in the group.

(3) Rounded indications are those having a length less than three times the width and may be circular, elliptical, or irregular in shape. Permitted rounded indications shall be as shown in Table 3.5.2B.

(4) For reinforcing bar, the acceptance criteria shall be as shown in Table 3.5.2C.

3.5.3 Bend Specimens

3.5.3.1 Groove Welds. The location and number of bend specimens are shown in 3.7. The preparation of specimens and guided bend test fixture requirements are specified in Appendix A.

(1) After bending, the center of the weld shall be approximately in the center of the bent portion of the specimen, and each specimen shall be visually examined for evidence of defects. An acceptable test shall be one in which no specimen exhibits any crack or open defect exceeding 1/8 in. (3.2 mm) in any direction. Cracks occurring at the corner of specimens shall not be considered unless there is definite evidence that they result from slag inclusion or other internal defects.

3.5.3.2 Clad Welds. For corrosion resistant cladding, no open defect exceeding 1/16 in. (1.6 mm) measured in any direction on the surface of the cladding, and no open defects exceeding 1/8 in. (3.2 mm) are permitted at the weld interface after bending.

Table 3.5.2B
Acceptable rounded discontinuities

Weld metal thickness (t), in. (mm)	Max acceptable discontinuities	
	Single	Multiple
Less than 1/8 (3.2)	20% of t	A maximum of 12 acceptable indications in 6 in. (150 mm) of weld.*
1/8 (3.2) or thicker**	The lesser of 20% of t or 1/8 in. (3.2 mm)	Per Chart Appendix D

*For welds less than 6 in. (150 mm) in length, a proportionally fewer number of indications shall be permitted.

**Rounded indications less than 1/32 in. (0.8 mm) maximum diameter shall not be considered in this thickness range. Root concavity is permitted provided the film density through the area of interest is not less than that through the base metal.

Table 3.5.2C
Acceptable rebar discontinuities

Bar size	Sum of discontinuity dimensions		Single discontinuity dimensions	
	in.	mm	in.	mm
8-9	3/16	4.8	1/8	3.2
10	1/4	6.4	1/8	3.2
11	1/4	6.4	3/16	4.8
14	5/16	8.0	3/16	4.8
18	7/16	11.2	1/4	6.4

3.5.4 Bend/Break Tests - Fillet Welds

3.5.4.1 The location of bend/break specimens is shown in 3.7.3. The fillet weld bend/break specimen, (Figs. 3.7.3A, B, C, D or E), shall be bent with the root in tension until the specimen either fractures or until it is bent flat upon itself. The specimen shall pass if it does not fracture, or if the fillet fractures, the fractured surface shall show complete fusion to the joint root and shall exhibit no inclusion or porosity larger than 3/32 in. (2.4 mm) in its greatest dimension.

3.5.5 Macro Examination (Appendix E)

3.5.5.1 The location and number of macroetch specimens are shown in Figs. 3.7.3, 3.7.4, and 3.7.5.

(1) Examination of macroetch specimens for the standard fillet test weldments and workmanship weldments shall show complete fusion and freedom from cracks. Other indications at the root, not exceed-ing 1/32 in. (0.8 mm) shall be acceptable. Concavity or convexity of the weld face shall not exceed 1/16 in. (1.6 mm). Fillet legs shall not differ in size by more than 1/8 in. (3.2 mm).

(2) Examination of reinforcing bar test weldment macroetch specimens in 3.7.4 shall show complete fusion for butt joint specimens and the required throat for flare groove weld specimens. There shall be no cracks or incomplete fusion evident in the weld metal or heat-affected zone.

(3) Examination of hardfacing overlays shall show complete fusion.

3.5.6 Penetrant Examination (Appendix F)

3.5.6.1 For weld cladding shown in 3.7.5, the entire surface of the test weldment shall be penetrant examined at the minimum qualified thickness in accordance with Fig. A3B (Appendix F). To pass the test, there shall be:

(1) No relevant linear indications with surface dimensions greater than 1/16 in. (1.6 mm). A linear indication is one in which the length is more than 3 times the width.

(2) No more than four rounded indications of any size, in a line separated by 1/16 in. (1.6 mm) or less, except where the WPS specifies either more or less stringent requirements.

3.5.7 Tension Test - Reinforcing Bar.
The tensile strength shall not be less than 125 percent of the minimum specified yield strength of the type and grade of bars joined (of the lower specified yield strength for dissimilar material joints).

3.6 Welding Variables Affecting Performance Qualification

3.6.1 Welders and welding operators who qualify for one WPS are also qualified to weld with any other WPS using the same process within the limits of performance welding variables.

3.6.2 Performance Welding Variables. In addition to the tables in 3.4, the following require requalification of a welder or welding operator.

3.6.2.1 Welders

(1) A change in welding process.

(2) The addition or deletion of backing (including weld metal made with another process) in single-welded groove joints.

(3) A change in the weld metal thickness or pipe diameter range beyond that for which the welder is qualified.

(4) A change in filler metal F Number except as allowed in 3.4.4.

(5) A change in welding position except as allowed by 3.4.6.

(6) For OFW, a change in the type of fuel gas.

(7) For GTAW, a change from ac to dc, or vice versa, or a change in polarity.

(8) A change in vertical weld progression (upward or downward).

(9) For GMAW, a change from spray arc, globular arc, or pulsed arc to short-circuiting arc, or vice versa.

(10) For GTAW or PAW, the omission or addition of consumable inserts.

(11) For GTAW, the deletion of root shielding gas except for double welded butt joints and fillet welds.

3.6.2.2 Welding Operators

(1) A change in welding process.

(2) A change in welding position except as allowed by 3.4.6.

3.7 Standard Test Weldments and Test Specimen Location

3.7.1 Standard Pipe Groove Test Weldment. The standard pipe groove test weldment for performance qualification shall consist of two pipe sections, each a minimum of 3 in. (76 mm) long joined by welding to make one test weldment a minimum of 6 in. (150 mm) long. The diameter and wall thickness shall be based upon the requirements of Table 3.4.5A. See Figs. 3.7.1A through 3.7.1C.

3.7.1.1 Required Specimens. For qualification in the 1FR or 2G positions the following specimens are required. The specimens are to be removed approximately 180 deg. apart.

　　1 Face Bend - Note 1, Fig. 3.7.1B
　　1 Root Bend - Note 1, Fig. 3.7.1B

3.7.1.2 Required Specimens (see Fig. 3.7.1B). For qualification in the 5G, 6G, or 6GR positions, the following specimens are required:

　　2 Face Bends - Note, Fig. 3.7.1B
　　2 Root Bends - Note, Fig. 3.7.1B

3.7.1.3 Required Specimens (see Fig. 3.7.1C). For qualification of the 2G + 5G positions on a single pipe weld the following specimens are required:

2G position:

　　1 Face Bend - Note 1, Fig. 3.7.1C
　　1 Root Bend - Note 1, Fig. 3.7.1C

5G position:

　　2 Face Bends - Note 1, Fig. 3.7.1C
　　2 Root Bends - Note 1, Fig. 3.7.1C

3.7.2 Standard Plate Groove Test Weldment. The standard plate groove performance test weldments are shown in Figs. 3.7.2A and 3.7.2B. Minimum length of test weldment shall be based upon the type of specimens required. Plate thickness shall be as shown in Table 3.4.5B.

3.7.2.1 Required Transverse Bend Specimens (see Fig. 3.7.2A):

　　Face Bend - Note 1, Fig. 3.7.2A
　　Root Bend - Note 1, Fig. 3.7.2A

3.7.2.2 Required Longitudinal Specimens (see Fig. 3.7.2B):

　　1 Face Bend
　　1 Root Bend

3.7.3 Standard and Alternate Fillet Welds. The standard and alternate fillet weld performance test weldment shall be essentially as shown in Figs. 3.7.3A, B, C, D, or E and test specimens shall be removed as shown. The thickness of the test weldment shall be based upon the requirements of Table 3.4.5C.

3.7.3.1 Required Specimens - Plate:

　　1 Bend/Break Test
　　2 Macroetch

3.7.3.2 Required Specimens - Pipe:

　　1 Bend/Break Test
　　2 Macroetch

3.7.4 Reinforcing Bar Test Weldments. For each performance test, two weldments shall be made as

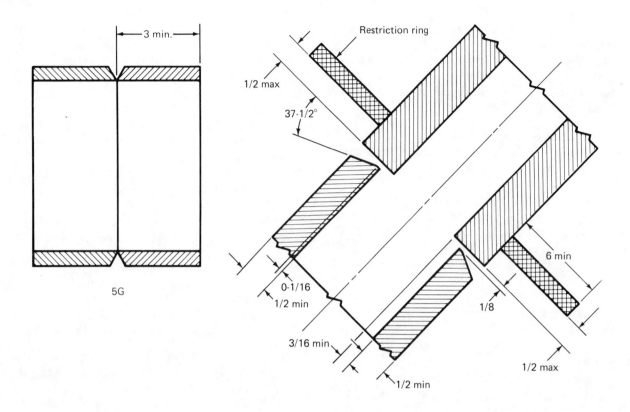

Fig. 3.7.1A — Standard 5G, and 6GR pipe test weldments

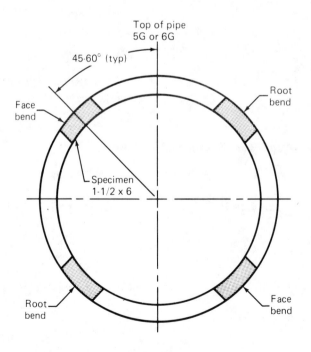

Fig. 3.7.1B — Specimen location

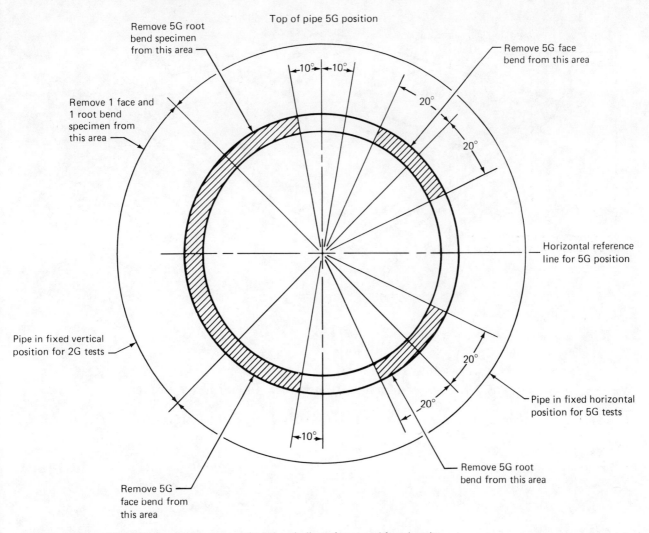

Remove 5G root bend specimen from this area

Top of pipe 5G position

Remove 5G face bend from this area

Remove 1 face and 1 root bend specimen from this area

Horizontal reference line for 5G position

Pipe in fixed vertical position for 2G tests

Pipe in fixed horizontal position for 5G tests

Remove 5G face bend from this area

Remove 5G root bend from this area

Note: For pipe over 3/8 in. thick, side bends may be taken in lieu of root and face bends.

Fig. 3.7.1C — Specimen location for 2G and 5G positions

shown in Figs. 3.7.4A or 3.7.4B, as applicable, and subjected to the following tests:

3.7.4.1 Required Specimens:
1 Full Section Tension Test
1 Macroetch Test

3.7.5 Surfacing Test Weldments. Weld cladding and hardfacing.

3.7.5.1 Required Specimens for Cladding (see Fig. 3.7.5A):
Penetrant Examination of Clad Surface
2 Transverse Side Bends

3.7.5.2 Required Specimens for Hardfacing (see Fig. 3.7.5B):
3 Hardness Readings - From the minimum thickness qualified
Macroetch Examination (2 Faces)

3.7.6 Workmanship Test Weldments. The examples in Fig. 3.7.6 illustrate typical workmanship samples. The thicknesses and configuration of the test weldment shall be as specified by the qualifier. These examples illustrate how a typical test weldment might appear.

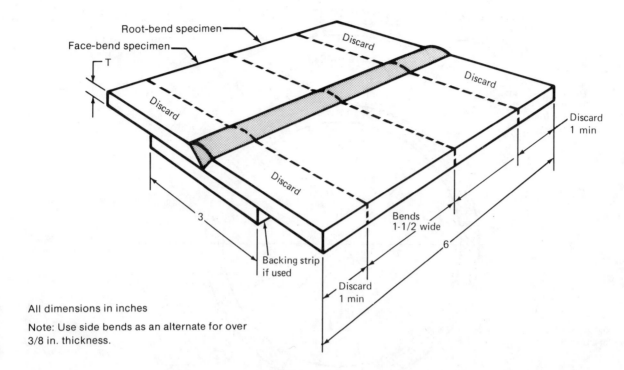

All dimensions in inches

Note: Use side bends as an alternate for over 3/8 in. thickness.

Fig. 3.7.2A — Standard plate transverse bend test weldments

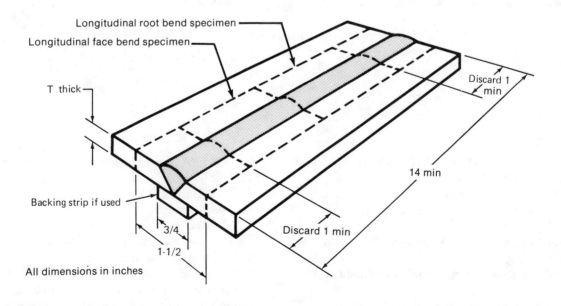

All dimensions in inches

Fig. 3.7.2B — Alternate plate test weldment longitudinal bends (see 2.8.2.3)

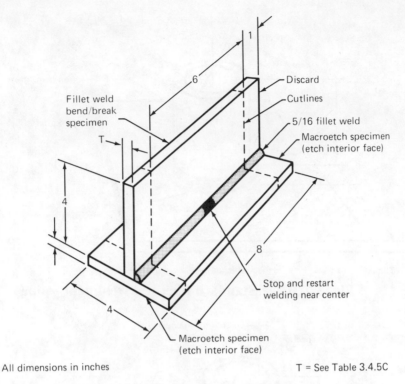

All dimensions in inches   T = See Table 3.4.5C

Note: Plate thickness and dimensions are minimum.

Fig. 3.7.3A — Standard fillet weld test weldment - plate

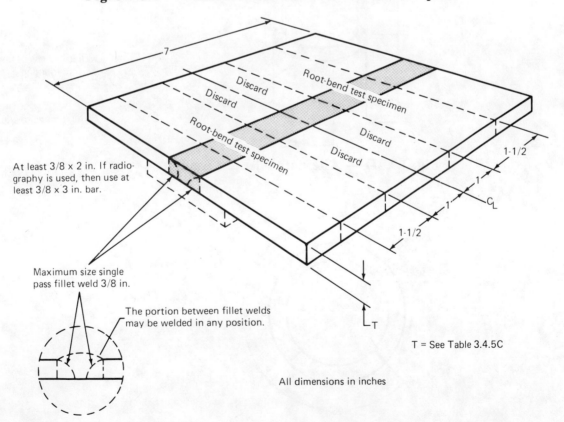

T = See Table 3.4.5C

All dimensions in inches

Fig. 3.7.3B — Alternate fillet weld test weldment - plate

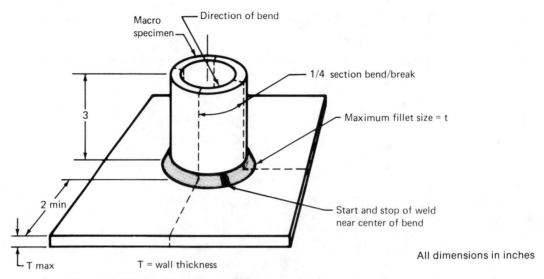

Fig. 3.7.3C — Standard fillet test weldment - pipe

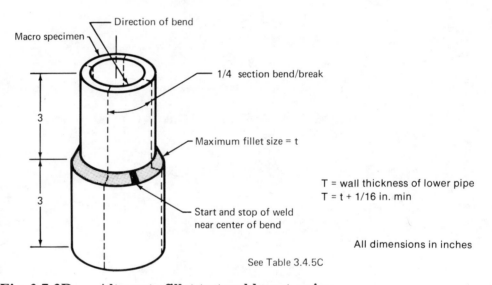

Fig. 3.7.3D — Alternate fillet test weldment - pipe

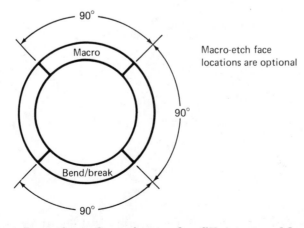

Fig. 3.7.3E — Location of specimens for fillet test weldments - pipe

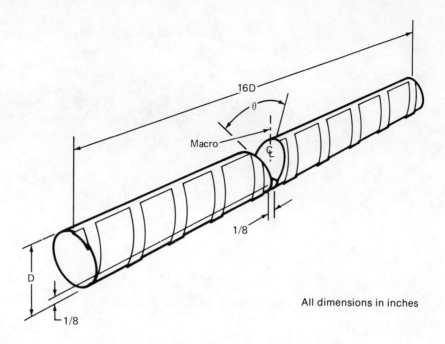

16D

θ

Macro

℄

1/8

D

1/8

All dimensions in inches

(1) For bars No. 9 or larger, use single-V or bevel-groove weld (θ=45 to 60)
(2) For bars No. 8 or smaller, use single-V with split-pipe backing (θ=60)

**Fig. 3.7.4A — Butt joint test weldment - reinforcing bar
Complete joint penetration groove weld**

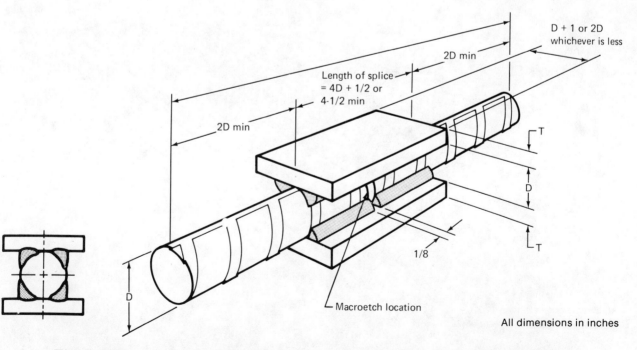

D + 1 or 2D
whichever is less

2D min

Length of splice
= 4D + 1/2 or
4-1/2 min

2D min

T

D

T

1/8

D

Macroetch location

All dimensions in inches

Fig. 3.7.4B — Spliced butt joint (flare groove) test weldment - reinforcing bar

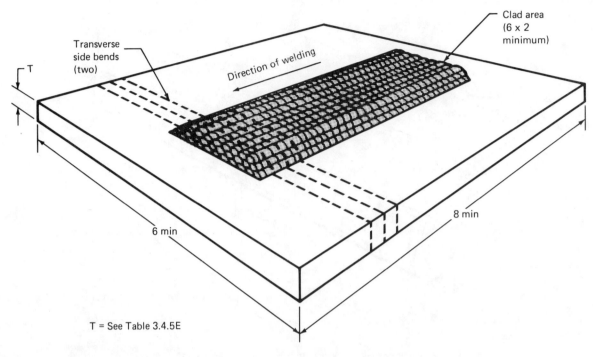

All dimensions in inches

Note: Weld layers are illustrative only.

Fig. 3.7.5A — Location of cladding specimens

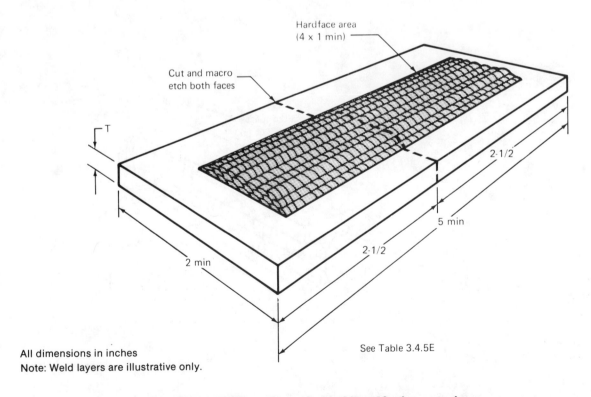

All dimensions in inches
Note: Weld layers are illustrative only.

Fig. 3.7.5B — Location of hardfacing specimens

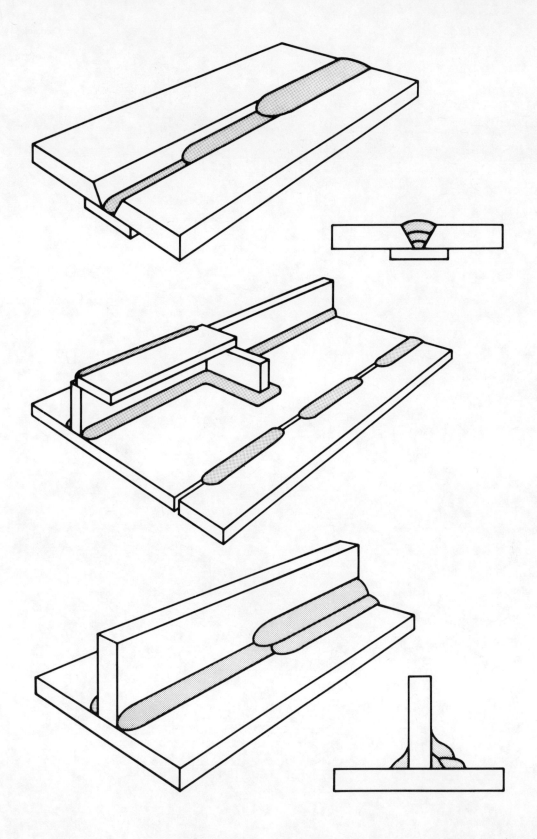

Fig. 3.7.6 — Workmanship test weldments

Appendix A
Illustrations - Weld Position, Test Specimens,
Test Jigs, and Suggested Forms

A1. Welding Positions - Illustrated

A1.1 Groove Weld Position Changes. Figure A1.1A specifies the points at which groove weld position changes occur.

A1.2 Fillet Weld Position Changes. Figure A1.1B specifies the points at which fillet weld position changes occur.

A1.3 Standard Groove Test Weldment Positions. Figure A1.2A illustrates and identifies the positions permitted for Standard Groove Test Weldments. Standard Groove Test Weldments shall be positioned in the applicable position(s) as shown in Fig. A1.2A, except that an angular deviation is allowed from the true horizontal or vertical plane in accordance with Fig. A1.1A.

A1.4 Standard Fillet Test Weldment Positions. Figure A1.2B illustrates and identifies the positions for Standard Fillet Weld Test Weldments. Standard Fillet Test Weldments shall be positioned in the applicable position(s) as illustrated.

A1.5 Standard Stud Test Weldment Positions. Figure A1.3A illustrates and identifies the positions for Standard Stud Test Weldments. Standard Stud Test Weldments shall be positioned in the applicable position illustrated within the limitations shown in Fig. A1.3B.

A2. Guided Bend Specimens

A2.1 Preparation of Groove Weld Specimens. Guided bend test specimens shall be prepared by cutting the test plate or pipe to form specimens as illustrated in Figs. A2.1A, A2.1B, and A2.1C. The cut surfaces are designated the specimen sides. The other two surfaces are designated the face and root surfaces.

A2.2 Weld reinforcement and backing shall be removed flush with the specimen surface. Cut surfaces shall be parallel, may be thermally cut, and shall be machined or ground a minimum of 1/8 in. (3.2 mm) on thermally cut edges, except that M-1 metals may be bent "as-cut" if no objectionable surface roughness exists.

A2.3 Subsize Transverse Face and Root Bends. For pipe of 4 in. (100 mm) outside diameter or less, the bend specimen width may be 3/4 in. (19 mm), measured around the outside surface. Alternatively, for outside diameters less than 2-7/8 in. (73 mm), the width may be that obtained by cutting the pipe into quarter sections.

A2.4 Nonstandard Bend Specimens. For base metal thickness less than 3/8 in. (9.5 mm), the thickness of the specimen may be the thickness of the base metal, except that M-23 and M-35 materials [excluding alloys C95200 and C95400 for which 3/8 in. (9.5 mm) is required] shall be a maximum of 1/8 in. (3.2 mm) thick. For metals less than 1/8 in. (3.2 mm) thick, the specimen thickness shall be the thickness of the base metal.

A3. Surfacing Specimens

Chemical analysis and bend test specimens shall be prepared as shown in Figs. A3A and A3B.

A4. Tension Specimens (See Figs. A4A-A4F)

A4.1 A single specimen may be used for thicknesses of 1 in. or less.

A4.2 For thicknesses over 1 in., single or multiple specimens may be used [except as permitted by A4.2(3) provided (1) and (2) are complied with].

(1) Collectively, multiple specimens, representing the full thickness of the weld at one location, shall comprise a set.

(2) The entire thickness shall be mechanically cut into approximately equal strips. For specimens that are not turned, specimen thicknesses shall be the maximum size that can be tested in available equipment.

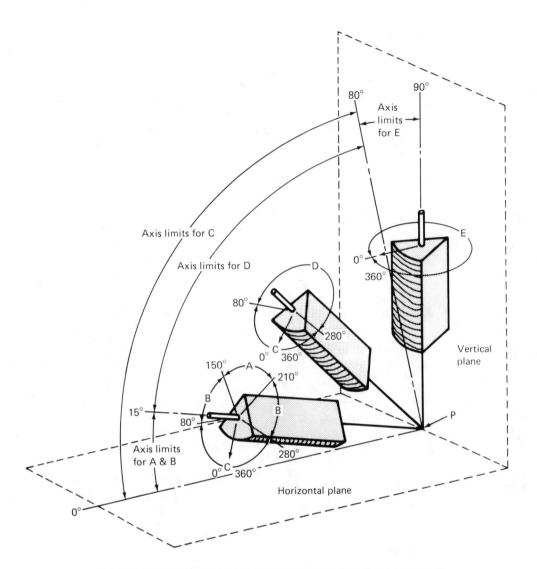

Tabulation of positions of groove welds

Position	Diagram reference	Inclination of axis	Rotation of face
Flat	A	0° to 15°	150° to 210°
Horizontal	B	0° to 15°	80 to 150 210° to 280°
Overhead	C	0° to 80°	0° to 80° 280° to 360°
Vertical	D E	15° to 80° 80° to 90°	80° to 280° 0° to 360°

Notes:
1. The horizontal reference plane is always taken to lie below the weld under consideration.
2. The inclination of axis is measured from the horizontal reference plane toward the vertical reference plane.
3. The angle of rotation of the face is determined by a line perpendicular to the theoretical face of the weld which passes through the axis of the weld. The reference position (0°) of rotation of the face invariably points in the direction opposite to that in which the axis angle increases. When looking at point P, the angle of rotation of the face of the weld is measured in a clockwise direction from the reference position (0°).

Fig. A1.1A — Position of groove welds

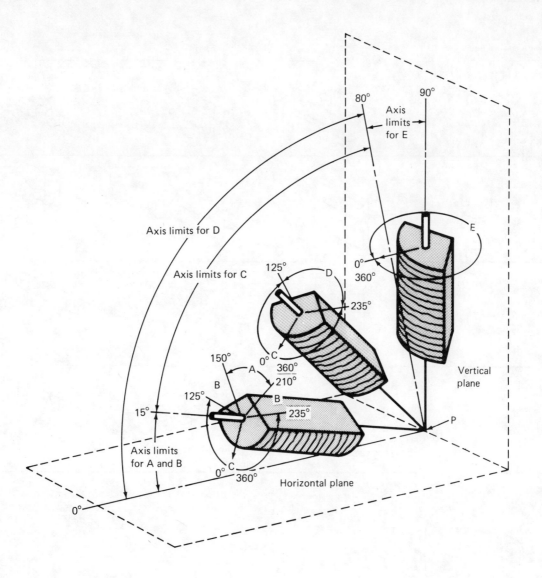

Tabulations of positions of fillet welds			
Position	Diagram reference	Inclination of axis	Rotation of face
Flat	A	0° to 15°	150° to 210°
Horizontal	B	0° to 15°	125° to 150° 210° to 235°
Overhead	C	0° to 80°	0° to 125° 235° to 360°
Vertical	D E	15° to 80° 80° to 90°	125° to 235° 0° to 360°

Fig. A1.1B — Position of fillet welds

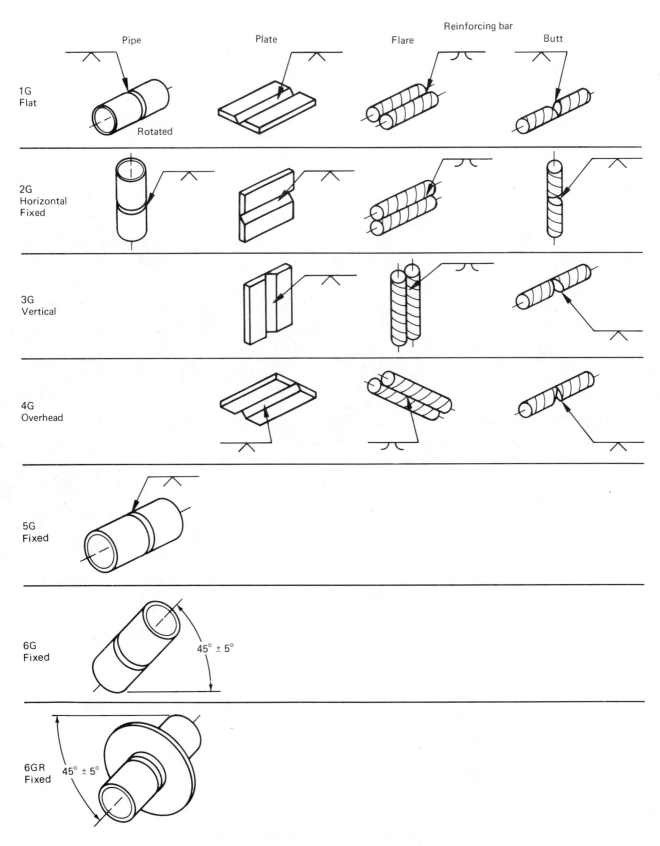

Fig. A1.2A — Standard groove test weldment positions

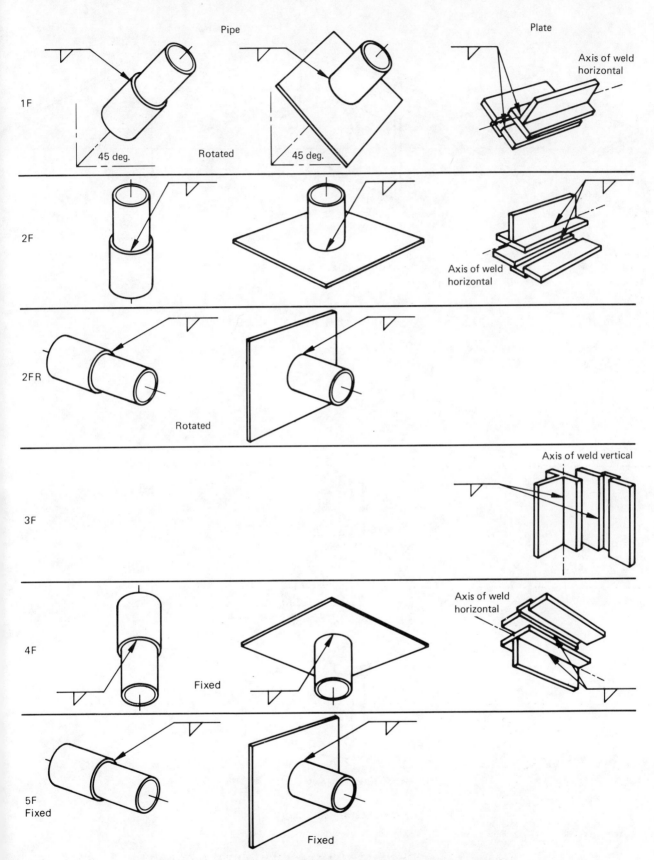

Fig. A1.2B — Standard fillet weldment test positions

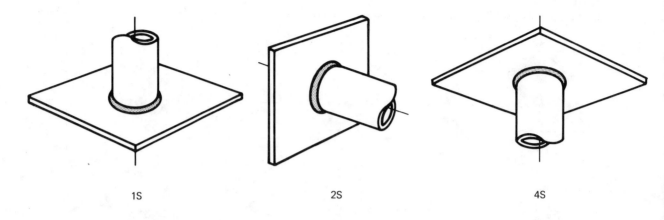

STUD WELDING POSITIONS

Fig. A1.3A — Standard stud welding positions

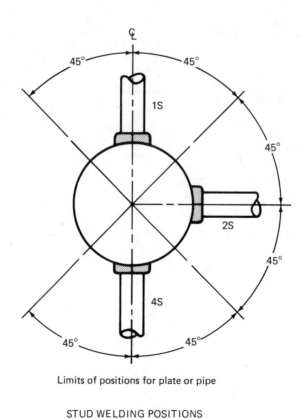

Limits of positions for plate or pipe

STUD WELDING POSITIONS

Fig. A1.3B — Stud welding position limitations

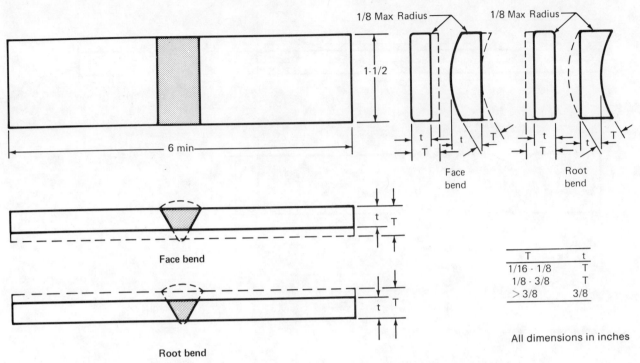

Fig. A2.1A — Transverse face and root bend specimens

T	t
1/16 - 1/8	T
1/8 - 3/8	T
> 3/8	3/8

All dimensions in inches

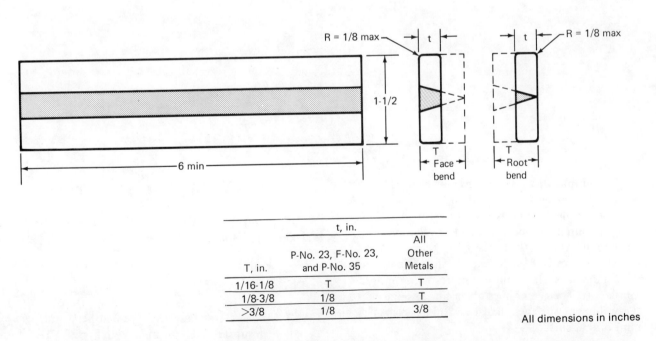

T, in.	t, in.	
	P-No. 23, F-No. 23, and P-No. 35	All Other Metals
1/16-1/8	T	T
1/8-3/8	1/8	T
>3/8	1/8	3/8

All dimensions in inches

Note: A longer specimen length may be necessary when using a wraparound type bending fixture or when testing steel with a yield strength of 90 ksi or more.

Fig. A2.1B — Longitudinal face and root bend specimens

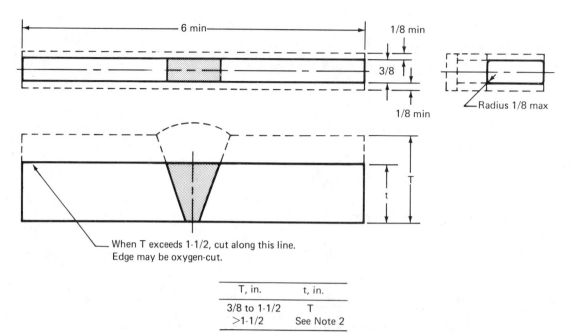

All dimensions in inches

Notes:

1. A longer specimen length may be necessary when using a wraparound-type bending fixture or when testing steel with a yield point of 90 ksi or more.
2. For plates over 1-1/2 in. thick, cut the specimen into approximately equal strips with t between 3/4 and 1-1/2 in. and test each strip.

Fig. A2.1C — Transverse side bend specimens

(3) Multiple turned specimens may be used for test weldment thicknesses greater than 1 in. (25.4 mm). A set of multiple specimens shall be used to satisfy a single required tension test. Collectively, specimens representing the full thickness of the weld at one location shall comprise a set. Specimens shall be parallel to the weldment surface and not more than 1 in. (25.4 mm) apart. The center lines of specimens adjacent to the weldment surfaces shall be within 5/8 in. (15.9 mm) of that surface.

A4.3 Cold straightening of test specimens is permitted prior to removal of reinforcement.

A4.4 Weld reinforcement shall be removed flush to the base metal.

A5. Fillet Weld - Shear Specimens

Fillet shear test specimens shall be prepared in accordance with Fig. A5.

A6. Test Jigs

A6.1 Guided Bend Test Jigs. Guided bend test jigs shall be in substantial agreement with Figs. A6.1A or A6.1B (for construction details, see AWS B4.0, *Standard Methods for Mechanical Testing of Welds*).

A6.2 Stud Weld Test Jigs. The jig for the stud weld bend test shall be substantially as shown in Fig.

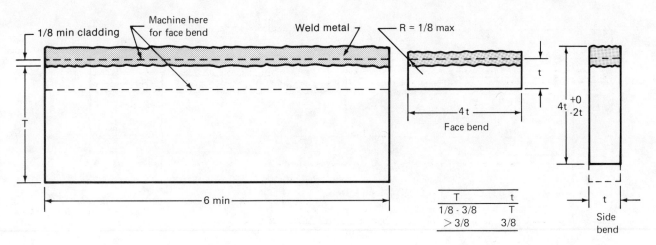

All dimensions in inches

Notes:

1. The minimum amount necessary shall be machined from the face-bend weld cladding surface to obtain a smooth surface.
2. T = the thickness of base metal.

Fig. A3A — Weld cladding side and face bend specimens

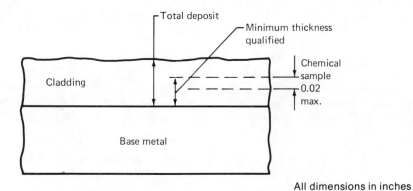

All dimensions in inches

Fig. A3B — Weld cladding and hardfacing chemical analysis specimens

A6.2A. The jig for the torque testing of threaded studs is shown in Fig. A6.2B, and the jig for tension testing of studs is shown in Fig. A6.2C.

A7. Suggested Sample Welding Forms

A7.1 Welding Procedure Specification (WPS) Form. Form A7.1 is a suggested form for recording the essen-tial information needed in preparing a WPS.

A7.2 Procedure Qualification Record (PQR) Form. Form A7.2 is a suggested form for recording the essen-tial information and test results for a PQR.

A7.3 Performance Qualification Test Record Form. Form A7.3 is a suggested form for recording the essen-tial information to substantiate the performance qual-ification of a welder or welding operator.

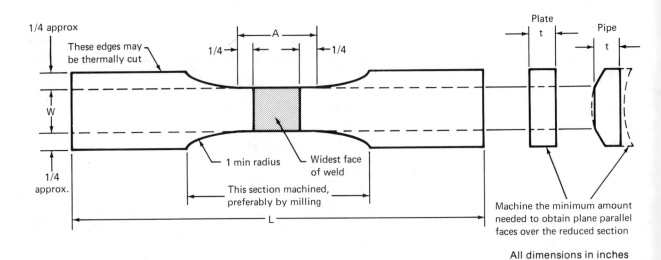

All dimensions in inches

	Dimensions			
	Plate test		Pipe test	
				6 in. & 8 in.
			2 in. & 3 in.	diameter or large
	T ≤ 1 in.	T ≥ 1 in.	diameter	job size pipe
A - Length of reduced section, in.	Widest face of weld + 1/2, 2-1/4 min		Widest face of weld + 1/2 in., 2-1/4 min	
L - Overall length, approximate	10″ or as required by testing equipment		10″ or as required by testing equipment	
W - Approximate, in.	1-1/2	1	1/2	3/4
t - Specimen thickness	T	T	Maximum possible with plane parallel faces within length A	

Notes:

1. T = thickness of the plate.

2. The ends of the reduced section shall not differ in width by more than 0.004 in. There may be a gradual decrease in width from the ends to the center, but the width at either end shall not be more than 0.015 in. larger than the width at the center.

Fig. A4A — Reduced section tension specimens

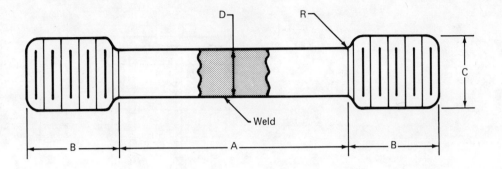

	Standard Dimension			
	(a)	(b)	(c)	(d)
	0.505 Specimen	0.353 Specimen	0.252 Specimen	0.188 Specimen
A - Length, reduced section, in.	See Note 4	See Note 4	See Note 4	See Note 4
D - Diameter, in.	0.500 ± 0.010	0.350 ± 0.007	0.250 ± 0.005	0.188 ± 0.003
R - Radius of fillet, in.	3/8, min.	1/4, min.	3/16, min.	1/8, min.
B - Length of end section, in.	1-3/8, approx.	1-1/8, approx.	7/8, approx.	1/2, approx.
C - Diameter of end section, in.	3/4	1/2	3/8	1/4

Notes:

1. Use maximum diameter specimen (a), (b), (c), or (d) that can be cut from the section.
2. Weld should be in center of reduced section.
3. Where only a single specimen is required, the center of the specimen should be midway between the surfaces.
4. Reduced Section "A" should not be less than width of weld plus two "D."
5. The ends may be of any shape to fit the holders of the testing machine in such a way that the load is applied axially.

Fig. A4B — Tension reduced section-turned specimens

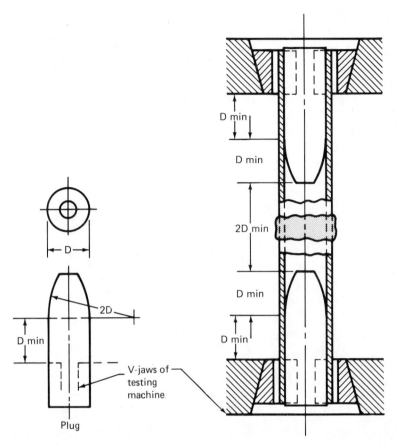

Fig. A4C — Full section tension alternate specimen for pipe 2 in. (nominal) or smaller

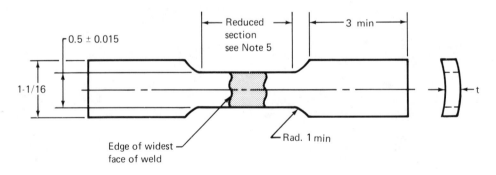

All dimensions in inches

Notes:

1. Cross-sectional area = 0.5 x T.
2. Test specimen thickness (t) shall be within the material thickness range allowed by the applicable material specification for the schedule/diameter of pipe being tested.
3. The specimen reduced section shall be parallel with 0.010 in. Specimen width may be gradually tapered provided the ends are no more than 0.010 in. wider than the center.
4. Weld reinforcement shall be removed so that weld thickness does not exceed the base metal thickness.
5. The reduced section shall not be less than the width of the welds plus 2t, and shall be machined preferably by milling.

Fig. A4D — Reduced section tension alternate specimen for pipe 3 in. or smaller

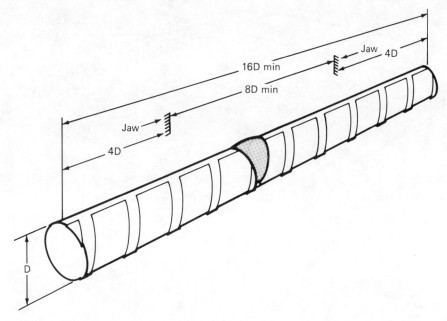

Fig. A4E — Butt joint specimen for full section tension test - reinforcing bar

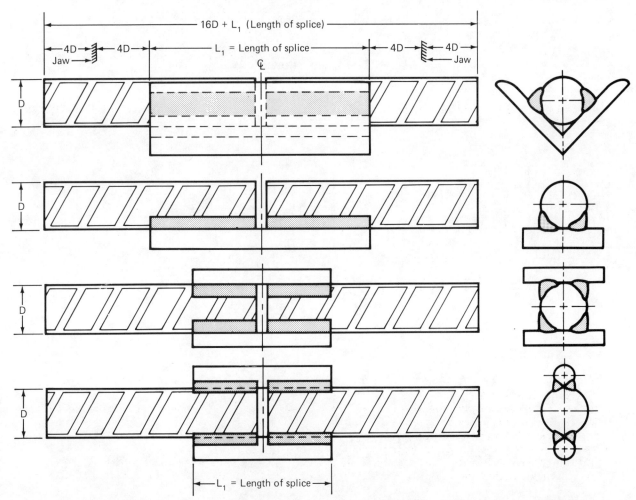

Fig. A4F — Spliced butt joint (flare groove) specimen for full section tension test

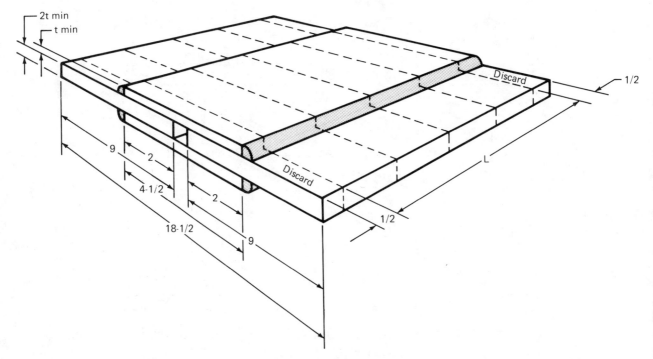

L = Length should be sufficient for the required number of specimens, which may be of any convenient width not less than 1 in.

t = Specified fillet weld size plus 1/8

Fig. A5 — Shear specimens

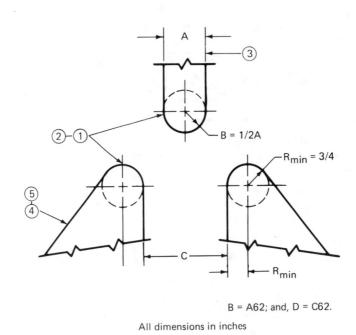

$B = A62$; and, $D = C62$.

All dimensions in inches

Fig. A6.1A — Guided bend test jigs

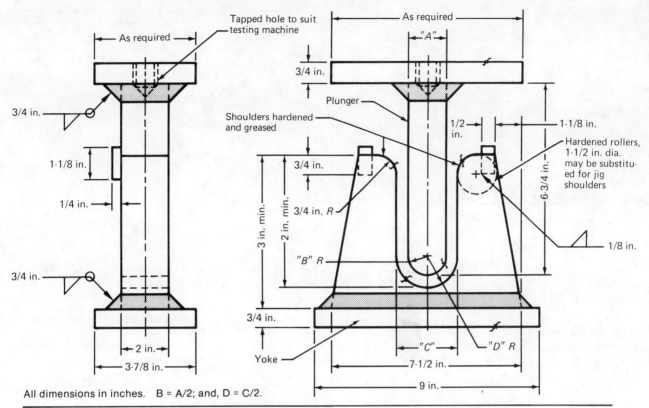

All dimensions in inches. B = A/2; and, D = C/2.

Material	Specimen thickness in. (mm)	A(t or in. [mm])	C(t or in. [mm])
M-2X, welded with F-23	—	—	—
M-23, as welded	1/8 (3.2)	2-1/16 (52.4)	2-3/8 (60.3)
M-35, except as shown below	Less than 1/8	16-1/2t	18-1/2 t + 1/16 (1.6)
M-11			
M-23, annealed	3/8 (9.5)	2-1/2 (63.5)	3-3/8 (85.7)
M-25	Less than 3/8 (9.5)	6-2/3t	8-2/3 t + 1/8 (3.2)
M-35, SB-148 and SB-271, Alloys CDA 952 and 954			
M-51	1/16 (1.6) to 3/8 (9.5) (inclusive)	8 t	10 t + 1/8
M-51	1/16 (1.6) to 3/8 (9.5) (inclusive)	10 t	12 t + 1/8
M-61			
All others	3/8 (9.5)	1-1/2 (380)	2-3/8 (60.3)
	Less than 3/8 (9.5)	4 t	6 t + 1/8 (3.2)

Notes:
1. Either hardened and greased shoulders or hardened rollers free to rotate shall be used.
2. Shoulders or rollers shall have a minimum bearing surface of 2 in. and be high enough above the bottom of the jig so that specimens will clear the rollers when the ram is in the low position.
3. The ram and base shall be sufficiently rigid to prevent deflection and misalignment while making the bend test. The body of the ram may be less than the dimensions shown in Column A.
4. Roller supports shall be designed to minimize deflection or misalignment and equipped with means for maintaining the rollers centered, midpoint, and aligned with respect to the ram.

Fig. A6.1A (Continued) — Guided bend test jigs

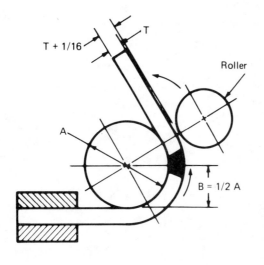

Material	Thickness of specimens (t) in.	Dimension "A" in.
M-52	1/16 - 3/8 incl.	10t
M-51 and M-61	1/16 - 3/8 incl.	8 t
M-11 and M-25	3/8 t	2-1/2 6-2/3 t
M-23 and M-35 B-171, Alloy 628	1/8 —	2-1/16 —
All others	0.0299 - 0.1345 3/8 t	1/2 1-1/2 4 t

Notes:

1. Dimensions not shown are the option of the designer. The essential consideration is to have adequate rigidity so that the jig parts will not spring.

2. The specimen shall be firmly clamped on one end so that there is no sliding of the specimen during the bending operation.

3. Test specimens shall be removed from the jig when the outer roll has been moved 180 deg. from the starting point.

Fig. A6.1B — Alternate guided bend wrap-around test jig

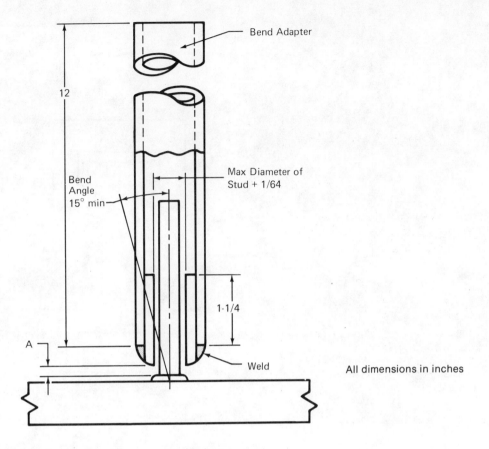

For Stud Diameter (in.)   1/8   3/16   1/4   3/8   1/2   5/8   3/4   7/8   1
Use Adapter Gap "A" (in.) 1/8   1/8   3/16   7/32   5/16   11/32   15/32   15/32   19/32

Fig. A6.2A — Stud weld bend jig

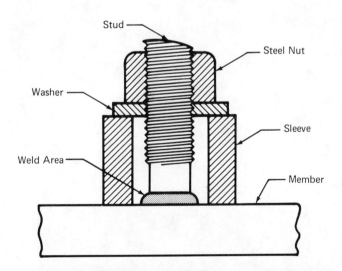

Dimensions are appropriate to the size of the stud.
Threads of the stud shall be clean and free of lubricant
other than residual cutting oil.

Fig. A6.2B — Torque testing arrangement for stud welds

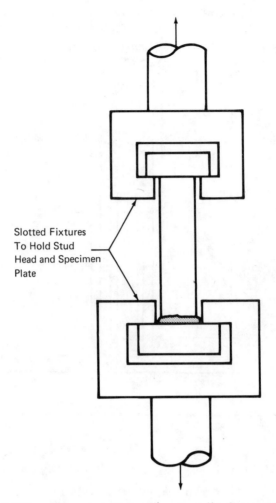

Slotted Fixtures
To Hold Stud
Head and Specimen
Plate

Fig. A6.2C — Suggested tension test jig for stud welds

Form A.7.1

SUGGESTED
WELDING PROCEDURE SPECIFICATION (WPS)

Identification _____

Date _____ Revision _____

Company name _____

Supporting PQR no.(s) _____ Type - Manual ( ) Semi-Automatic ( )

Welding process(es) _____ Machine ( ) Automatic ( )

Backing: Yes ( ) No ( )

Backing material (type) _____

Material number _____ Group _____ To material number _____ Group _____

Material spec. type and grade _____ To material spec. type and grade _____

Base metal thickness range: Groove _____ Fillet _____

Deposited weld metal thickness range _____

Filler metal F no. _____ A no. _____

Spec. no. (AWS) _____ Flux tradename _____

Electrode-flux (Class) _____ Type _____

Consumable insert: Yes ( ) No ( ) Classifications _____

Shape _____

Position(s) of joint _____ Size _____

Welding progression: Up ( ) Down ( ) Ferrite number (when reqd.) _____

PREHEAT: **GAS:**

 Preheat temp., min _____ Shielding gas(es) _____

 Interpass temp., max _____ Percent composition _____
 (continuous or special heating, where
 applicable, should be recorded) Flow rate _____
  Root shielding gas _____

POSTWELD HEAT TREATMENT: Trailing gas composition _____

 Temperature range _____ Trailing gas flow rate _____

 Time range _____

Tungsten electrode, type and size _____

Mode of metal transfer for GMAW: Short-circuiting ( ) Globular ( ) Spray ( )

Electrode wire feed speed range: _____

Stringer bead ( ) Weave bead ( ) Peening: Yes ( ) No ( )

Oscillation _____

Standoff distance

Multiple ( ) or single electrode ( )

Other _____

	Filler metal				Current			
Weld layer(s)	Process	Class	Dia.	Type & polarity	Amp range	Volt range	Travel speed range	
								e.g., Remarks, comments, hot wire addition, technique, torch angle, etc.

Approved for Production by _____
Employer

Note: Those items that are not applicable should be marked N.A.

Form A.7.2   **SUGGESTED**   **Page 1 of 2**
PROCEDURE QUALIFICATION RECORD (PQR)

WPS no. used for test _____   Welding process(es) _____
Company _____   Equipment type and model (sw) _____

JOINT DESIGN USED (2.6.1)

WELD INCREMENT SEQUENCE

Single ( )   Double weld ( )

POSTWELD HEAT TREAMTENT (2.6.6):
Backing material _____   Temp. _____
Root opening _____ Root face dimension _____   Time _____
Groove angle _____ Radius (J-U) _____   Other _____
Back gouging: Yes ( ) No ( ) Method _____

GAS (2.6.7)

BASE METALS (2.6.2)
   Gas type(s) _____
Material spec. _____ To _____   Gas mixture percentage
Type or grade _____ To _____   Flow rate _____
Material no. _____ To material no. _____   Backing gas _____ Flow rate _____
Group no. _____ To group no. _____   Root shielding gas
Thickness _____   EBW vacuum ( ) Absolute pressure ( )
Diameter (pipe) _____

ELECTRICAL CHARACTERISTICS (2.6.8)
Surfacing: Material _____ Thickness _____   Electrode extension _____
Chemical composition _____
Other _____   Standoff distance

FILLER METALS (2.6.3)
   Transfer mode (GMAW) _____
Weld metal analysis A no. _____   Electrode diameter tungsten _____
Filler metal F no. _____   Type tungsten electrode _____
AWS specification _____   Current: AC ( ) DCEP ( ) DCEN ( ) Pulsed ( )
AWS classification _____   Heat input _____
Flux class _____ Flux brand _____   EBW: beam focus current _____ Pulse freq. _____
Consumable insert: Spec. _____ Class. _____   Filament type _____ Shape ___ Size _____
Supplemental filler metal spec. _____ Class. _____   Other _____
Non-classified filler metals _____

TECHNIQUE (2.6.9)
Consumable guide (ESW) Yes ( ) No ( )   Oscillation frequency _____ Weave width _____
Supplemental deoxidant (EBW) _____   Dwell time _____

POSITION (2.6.4)
   String or weave bead _____ Weave width _____
Position of groove _____ Fillet _____   Multi-pass or single pass (per side) _____
Vertical progression: Up ( ) Down ( )   Number of electrodes _____
   Peening _____
   Electrode spacing _____

PREHEAT (2.6.5)
   Arc timing (SW) _____ Lift ( )
Preheat temp., actual min _____   PAW: Conventional ( ) Key hole ( )
Interpass temp., actual max _____   Interpass cleaning:

Pass no.	Filler metal size	Amps	Volts	Travel speed (ipm)	Filler metal wire (ipm)	Slope induction	Special notes (process, etc.)

Note: Those items that are not applicable should be marked N.A.

Form A.7.2   **Page 2 of 2**

TENSILE TEST SPECIMENS: SUGGESTED PROCEDURE QUALIFICATION RECORD   PQR No.

Type: _____ Tensile specimen size: _____ Area: _____

Groove ( )   Reinforcing bar ( )   Stud welds ( )

Tensile test results: (Minimum required UTS _____ psi)

Specimen no.	Width, in.	Thickness, in.	Area, in.^{2}	Max load lbs	UTS, psi	Type failure and location

GUIDED BEND TEST SPECIMENS - SPECIMEN SIZE: _____

Type	Result	Type	Result

MACRO-EXAMINATION RESULTS: Reinforcing bar ( )   Stud ( )

1. _____   4. _____
2. _____   5. _____
3. _____

SHEAR TEST RESULTS - FILLETS: 1. _____   3. _____
2. _____   4. _____

IMPACT TEST SPECIMENS

Type: _____ Size: _____

Test temperature: _____

Specimen location: WM = weld metal; BM = base metal; HAZ = heat-affected zone

Test results:

Welding position	Specimen location	Energy absorbed (ft.-lbs.)	Ductile fracture area (percent)	Lateral expansion (mils)

IF APPLICABLE   **RESULTS**

Hardness tests: ( ) Values _____   Acceptable ( )   Unacceptable ( )

Visual (special weldments 2.4.2) ( )   Acceptable ( )   Unacceptable ( )

Torque ( ) psi   Acceptable ( )   Unacceptable ( )

Proof test ( ) Method _____   Acceptable ( )   Unacceptable ( )

Chemical analysis ( )   Acceptable ( )   Unacceptable ( )

Non-destructive exam ( ) Process _____ Acceptable ( )   Unacceptable ( )

Other _____   Acceptable ( )   Unacceptable ( )

Mechanical Testing by (Company) _____ Lab No. _____

We certify that the statements in this Record are correct and that the test welds were prepared, welded, and tested in accordance with the requirements of the American Welding Society Standard for Welding Procedure and Performance Qualification (AWS B2.1-83).

Qualifier: _____   Reviewed by: _____

Date: _____   Approved by: _____
                                              Employer

Form A.7.3

SUGGESTED
PERFORMANCE QUALIFICATION TEST RECORD

Name _____ Identification _____ Welder ( )   Operator ( )

Social security number: _____ Qualified to WPS no. _____

Process(es) _____ Manual ( )   Semi-Automatic ( )   Automatic ( )   Machine ( )

Test base metal specification _____ To _____

Material number _____ To _____

Fuel gas (OFW) _____

AWS filler metal classification _____ F no. _____

Backing: Yes   ( )   No   ( )   Double (   ) or Single side (   )
Current: AC   ( )   DC   ( )   Short-circuiting arc (GMAW) Yes ( ) No ( )
Consumable insert: Yes   ( )   No   ( )
Root shielding: Yes   ( )   No   ( )

TEST WELDMENT	**POSITION TESTED**	**WELDMENT THICKNESS (T)**

GROOVE:

Pipe   1G ( )   2G ( )   5G ( )   6G ( )   6GR ( )   Diameter(s) _____ (T) _____
Plate   1G ( )   2G ( )   3G ( )   4G ( )   (T) _____
Rebar   1G ( )   2G ( )   3G ( )   4G ( )   Bar size _____ Butt   ( )
                                                          Spliced butt   ( )

FILLET:

Pipe ( )   1F ( )   2F ( )   3F ( )   4F ( )   5F ( )   Diameter _____ (T) _____
Plate ( )   1F ( )   2F ( )   3F ( )   4F ( )   (T) _____

Other (describe) _____

Test results: Remarks

Visual test results	N/A ( )	Pass ( )	Fail ( )
Bend test results	N/A ( )	Pass ( )	Fail ( )
Macro test results	N/A ( )	Pass ( )	Fail ( )
Tension test	N/A ( )	Pass ( )	Fail ( )
Radiographic test results	N/A ( )	Pass ( )	Fail ( )
Penetrant test	N/A ( )	Pass ( )	Fail ( )

QUALIFIED FOR:
PROCESSES
GROOVE: **THICKNESS**

Pipe   1G ( )   2G ( )   5G ( )   6G ( )   6GR ( )   (T) Min _____ Max _____ Dia _____
Plate   1G ( )   2G ( )   3G ( )   4G ( )   (T) Min _____ Max _____
Rebar   1G ( )   2G ( )   3G ( )   4G ( )   Bar size   Min _____ Max _____

FILLET:

Pipe   1F ( )   2F ( )   4F ( )   5F ( )   (T) Min _____ Max _____
Plate   1F ( )   2F ( )   3F ( )   4F ( )   (T) Min _____ Max _____
Rebar   1F ( )   2F ( )   3F ( )   4F ( )   Bar size   Min _____ Max _____

Weld cladding ( )   Position(s) _____ T Min _____ Max _____ Clad Min _____

Consumable insert ( )   Backing type ( )
Vertical Up ( )   Down ( )
Single side ( )   Double side ( )   No backing ( )
Short-circuiting arc ( )   Spray arc ( )   Pulsed arc ( )
Reinforcing bar - butt ( ) or Spliced butt ( )

The above named person is qualified for the welding process(es) used in this test within the limits of essential variables including materials and filler metal variables of the AWS Standard for Welding Procedure and Performance Qualification (AWS B2.1).

Date tested _____   Signed by _____
                                                        Qualifier

Appendix B
Filler Metal Classification

B1. F Numbers

The following F Number grouping of electrodes and welding rods (Table B1) is based primarily on their usability characteristics, which fundamentally determine the ability to make satisfactory welds with a given filler metal and process. This grouping is made to minimize the number of procedure and performance qualifications where this can logically be done. The grouping does not imply that filler metals within a group may be indiscriminately substituted for a metal which was used in the qualification test without consideration of the compatibility of the base and filler metals from the standpoint of metallurgical properties, postweld heat treatment, design, service requirements, and mechanical properties.

B2. A Numbers

The classification of ferrous weld metal analyses for procedure qualification is given in Table B2.

Table B1
F numbers - Grouping of electrodes and welding rods for qualification

F No.	AWS Specification Number	AWS Classification Number
		Steel and Steel Alloys
1	A5.1 & 5.5	EXX20, EXX24, EXX27, EXX28
2	A5.1 & 5.5	EXX12, EXX13, EXX14, EXXX13-X
3	A5.1 & 5.5	EXX10, EXX11, EXXX10-X, EXXX11-X
4	A5.1 & 5.5	EXX15, EXX16, EXX18, EXXX15-X, EXXX16-X, EXXX18-X, EXXX18-M
4	A5.4 Nom. Total Alloy 6% or less	EXXX15, EXXX16
4	A5.4 Nom. Total Alloy more than 6%	EXXX15, EXXX16
5	A5.4 Cr-Ni Electrode	EXXX15, EXXX16
7	A5.17 & A5.23	FXX-XXXX (FXX-EXXX), FXX-EXXX-X, FXX-ECXXX-X, FXX-EXXX-XN, and FXX-ECXXX-XN
8	A5.18 & A5.28	ERXXS-X, ERXXX-X, E-XXX-X
9	A5.9	ERXX
10	A5.20 & A5.29	EXXT-X
11	A5.22	EXXXT-X
12	A5.25	FESXX-EXXXX-EW, ENXX, FESXX-EXXXX
13	A5.26	EGXXTXX, EGXXSXXX

Table B1 - F numbers (Continued)

F No.	AWS Specification Number	AWS Classification Number
		Aluminum and Aluminum-Base Alloys
20	A5.3	A1-2, A1-43
21	A5.10	ER 1100
22	A5.10	ER 5554, ER 5356, ER 5556, ER 5183, ER 5654
23	A5.10	ER 4043, ER 4047
24	A5.10	R-SC 51A, R-SG 70A
		Copper and Copper-Base Alloys
31	A5.6 & 5.7	RCu, ECu
32	A5.6 & 5.7	RCuSi-A, ECuSi
33	A5.6 & 5.7	RCuSn-A, ECuSn-A, ECuSn-C
34	A5.6 & 5.7	RCuNi, ECuNi
35	A5.6 & 5.7	RBCu-Zn-A, RCuZn-C
36	A5.6 & 5.7	ER-CuA1-A1, ER-CuA1-A2, ECuA1-A2, ER-Cu-A1-A3, E-CuA1-B
37	A5.6 & 5.7	E-CuNiA1, E CuMnNiA1
38	A5.27	ERCuNiA1, ERCuMnNiA1
		Nickel and Nickel-Base Alloys
41	A5.11	E Ni-1
41	A5.14	ER Ni-1
42	A5.11	E NiCu-7
42	A5.14	ER NiCu-7
43	A5.11	E NiCrFe-1, 2, 3, 4, E NiCrMo-2, 3
43	A5.14	ER NiCr-3, ER NiCrFe-5, 6, ER NiCrMo-2, 3
44	A5.14	E NiMo-1, E NiCrMo-4, 5, ER NiMo-1, 2, 7 (Alloy B-2), ER NiCrMo-4, ER NiCrMo-5, ER NiCrMo-7 (Alloy C4)
45	A5.11	E NiCrMo-1
45	A5.14	ER NiCrMo-1, ER NiFeCr-1
		Titanium and Titanium Alloys
51	A5.16	ERTi-1, ERTi-2, ERTi-3, ERTi-4
		Zirconium and Zirconium Alloys
61	A5.24	ER Zr1, ER Zr2, ER Zr3, ER Zr4
		Magnesium Alloys
70	A5.19	ER AZ61A, ER AZ101A, ER AZ92A, ER AZ33A

Note: F No. 6 purposely omitted. F Nos. 7 through 13 listed in Table B1 are shown as F No. 6 in ASME IX.

Table B2
A Numbers — Classification of ferrous data analysis for procedure qualification

A No.	Types of weld deposit	Analyis*					
		C %	Cr %	Mo %	Ni %	Mn %	Si %
1	Mild Steel	0.15	—	—	—	1.60	1.00
2	Carbon-Moly	0.15	0.50	0.40-0.65	—	1.60	1.00
3	Chrome (0.4 to 2%)-Moly	0.15	0.40-2.00	0.40-0.65	—	1.60	1.00
4	Chrome (2 to 6%)-Moly	0.15	2.00-6.00	0.40-1.50	—	1.60	2.00
5	Chrome (6 to 10.5%)-Moly	0.15	6.00-10.50	0.40-1.50	—	1.20	2.00
6	Chrome-Martensitic	0.15	11.00-15.00	0.70	—	2.00	1.00
7	Chrome-Ferritic	0.15	11.00-30.00	1.00	—	1.00	3.00
8	Chromium-Nickel	0.15	14.50-30.00	4.00	7.50-15.00	2.50	1.00
9	Chromium-Nickel	0.30	25.00-30.00	4.00	15.00-37.00	2.50	1.00
10	Nickel to 4%	0.15	—	0.55	0.80- 4.00	1.70	1.00
11	Manganese-Moly	0.17	—	0.25-0.75	0.85	1.25-2.25	1.00
12	Nickel-Chrome-Moly	0.15	1.50	0.25-0.80	1.25- 2.80	0.75-2.25	1.00

*Single values shown are maximum.

Appendix C
Base Metal Groupings

C1. Table C1 indexes ASTM, API, and ASME metals in numerical order to provide ease of reference when determining into what grouping a particular metal specification falls.

C2. Table C2 groups base metals for welding procedure and performance qualification on the basis of mechanical properties, chemical composition, and metallurgical compatibility to minimize the number of welding qualifications where this can logically be done.

C2.1 Metal has been divided into general categories (e.g. M Numbers 1, 2, 3, 4 etc.) and further divided into groups within each general category.

C2.2 The category grouping does not imply that base metals may be substituted for other base metals within the same M Number without consideration for weldability.

C2.3 In addition to the material number and group number, a Specification column is provided which denotes the material specification, standard, or code which governs each listed metal. P denotes ASME materials, S denotes ASTM materials (exclusive of matching ASME specifications), and Q denotes listed materials which do not fall into either P or S categories.

Table C1
Numerical indexing

Base Metal Specification	Grade G- Class C- Type T-	Material No.	Group No.	Base Metal Specification	Grade G- Class C- Type T-	Material No.	Group No.
A31	G-A,B	1	1	A182	G-F1, F2	3	2
A36	---	1	1		G-F11, F11a, G-F11b, F12, F12b	4	1
A47	---	2C					
A48	C-20, 25, 30, 35, 40	2A			G-F21, F22a	5	1
	C-45, 50, 55	2B			G-F5, F5A, F7, F9	5	2
A53	T-S, G-A,B T-E, G-A,B T-F	1	1		G-F6aC1, G-F6aC4, G-F6B, G-F6NM	6	1
A105	---	1	2		G-F429	6	2
A106	G-A,B	1	1		G-F6a, G-Fb, C-1, C-2	6	3
	G-C	1	2		G-F430	7	2
A131	G-A,B, CS, D, E	1	1		G-F30's, F316's, F321's, F347's, F348's	8	1
	AH32AH36	1	2				
	DH323DH36	1	2				
	EH32EH36	1	2		G-F310, F10, F45	8	2
A135	G-A, B	1	1				
A139	G-A,B,C,D,E	1	1		G-FXM- 19, FXM-11	8	3
A159	G-1800, 2500, 3000, 3500, 4000	2A			G-F44, G-FR	8	4
A161	G-LowC,T1	1	1		G-FXM-27Cb	10-I	1
A167	T-301, 302, 302B, 304, 304L, 304LN, 305, 308, 309, 309S, 309Cb, 310, 310S, 310cb, 316, 316cb, 316Ti, 317, 317L, 321, 347, 348, XM-15	8	1	A192	---	1	1
				A197	---	2C	
				A199	G-T3b, T11	4	1
					G-T4, T21, T22	5	1
					G-T5, T7, T9	5	2
				A202	G-A, B	4	1
A178	G-A, C	1	1	A203	G-A, B	9A	1
A179	---	1	1		G-D, E, F	9B	1
A181	C-60	1	1	A204	G-A	3	1
	C-70	1	2		G-B, C	3	2
				A209	G-T1, T1a, T1b	3	1

Table C1 (Continued)

Base Metal Specification	Grade G-Class C-Type T-	Material No.	Group No.	Base Metal Specification	Grade G-Class C-Type T-	Material No.	Group No.
A210	G-A-1	1	1	A240	T-430, UNS 4400	7	2
	G-C	1	2	A240	T-302, 304, 304L, 304H, 304N, 316, 316LN, 316H, 316L, 316Cb, 316Ti, 316N, 317, 317L, 321, 321H, 347, 348, 347H, 348H, XM-15, XM-21	8	1
A213	G-T2	3	1				
	G-T3b, T11, T12	4	1				
	G-T-21, T-22	5	1				
	G-T5, T7, T9, T5b, T5c	5	2				
	G-TP304's, TP316's, TP321's, TP347, TP347H, TP348's, XM-15	8	1				
					T-309S, 310S, 309Cb, 310Cb	8	2
	G-TP310	8	2		T-XM-17, T-XM-18, T-XM-19, T-XM-29	8	3
	G-T17	10B	1				
A214	---	1	1				
A216	G-WCA	1	1		T-329, G-XM-27, T-XM-33	10I	9
	G-WCB, WCC	1	2				
A217	G-WC1	3	1		UNS 44700	10J	10
	G-WC4, WC5, WC6	4	1		UNS 44800	10K	11
	G-WC9	5	1	A242	T-1	1	1
	G-C5, C12	5	2		T-2	3	1
	G-CA15	6	3	A249	G-TP304's, TP316's, TP317, TP321's, TP347's, TP348's, TPXM-15	8	1
A225	G-C, D	10A	1				
A226	---	1	1				
A234	WPB	1	1				
	WPC	1	2				
	WP1	3	2				
	WP11, WP12	4	1		G-TP309, TP310	8	2
	WP22	5	1				
	WP5, WP7, WP9	5	2		G-TPXM-19, G-TPXM-29	8	3
	WPR	9A	1	A250	G-T1, T1a, T1b	3	1
A240	T-410	6	1	A266	C-1	1	1
	T-429	6	2		C-2, 4	1	2
	T-405, 409, 410's, XM-8, XM-27, XM-35	7	1				

Table C1 (Continued)

Base Metal Specification	Grade G- Class C- Type T-	Material No.	Group No.	Base Metal Specification	Grade G- Class C- Type T-	Material No.	Group No.
A268	G-TP410	6	1	A312	TP316L, TP316N, TP317, TP321, TP321H, TP347, TP347H, TP348, TP348H, TPXM-15	8	1
	G-TP429	6	2				
	G-TP405, 409	7	1				
	G-TP430, TPXM-8, T(18Cr-2Mo-Ti)	7	2				
	G-TP443	10D	1		G-TP309, TP310	8	2
	G-TP446, TP329	10E	1		G-TPXM-11, TPXM-19, TPXM-29	8	3
	G-TPXM-27, TPXM-33	10-I	9	A333	G-1 & 6	1	1
	UNS 44700	10-J	1		G-4	4	1
	UNS 44800	10-K	1		G-3	9B	1
A269	G-TP 304, 304L, 304LN, TP316, 316L, 316LN, TP317, TP321, TP348, TP347, G-TPXM-10, TPXM-11, TPXM-15, TPXM-19, TPXM-29	8	1		G-8	11A	1
				A334	G-1, 6	1	1
					G-7, 9	9A	1
					G-3	9B	1
					G-8	11A	1
				A335	G-P1, P2, P15	3	1
					G-P11, P12	4	1
					G-P21, P22	5	1
A270	T-304	8	1		G-P5, P5b, P5c, P7, P9	5	2
A271	G-TP304, TP304H, TP316, TP316H, TP321, TP321N, TP347, TP347H	8	1	A336	C-F1, C-F11, F-11A, F11B	3	2
					C-F12	4	1
A283	G-A,B,C,D	1	1		C-F21, F21A, F22, F22A	5	1
A284	G-C,D	1	1		C-5F, F5A, F9	5	2
A285	G-A,B,C	1	1		C-F6	6	3
A299	---	1	2		C-304s, 316s, 321s, 347s, 348s	8	1
A302	G-A	3	2		C-F310, C-FX11, FXM-19	8	2
	G-B, C,D	3	3				
A312	G-TP304, G-TP304H, TP304L, TP304N, TP316, TP316H,	8	1		C-FXM27Cb	10I	1
				A350	G-LF1	1	1
					G-LF2	1	2

Table C1 (Continued)

Base Metal Specification	Grade G- Class C- Type T-	Material No.	Group No.	Base Metal Specification	Grade G- Class C- Type T-	Material No.	Group No.
A350	G-LF3	9B	1	A381	All classes	1	1
	G-LF9	9A	1	A387	G-2 C1	3	1
A351	G-CF3, CF3A, CF8, CF8A, CF3M, CF8M, CG8M, CF8C, CF10, CF10M	8	1		G-2 C2	3	2
					G-11 C1, G-11 C2, G-12 C1, G-12 C2	4	1
	G-CH8, CH20, CK20, CN7M	8	2		G-21 C1, G-21 C2, G-22 C2	5	1
	G-CG6MMN	8	3		G-5 C1, G-5 C2	5	2
A352	G-LCA, LCB	1	1				
	G-LCC	1	2	A389	G-C23, C24	4	1
	G-LC1	3	1	A403	WP304, WP304H, WP304HF, WP304L, WP304N, WP316, WP316H, WP316L, WP316HF, WP316N, WP317, WP317L, WP321, WP321H, WP321HF, WP347, WP347H, WP347HF, WP348, WP348H	8	1
	G-LC2	9A	1				
	G-LC3	9B	1				
A353	---	11A	1				
A358	G-304, 304H, 304L, 304N, 316, 316L, 316H, 316N, 321, 347, 348	8	1				
	G-309, 310, G-XM-19	8	2				
	XM-29	8	3				
A369	G-FPA, FPB	1	1				
	G-FP1, FP2	3	1				
	G-FP3B, FP11, FP12	4	1				
	G-FP21, FP22	5	1		WP309, WP310	8	2
	G-FP5, FP7, FP9	5	2		XM-19	8	3
A372	T-I	1	1	A405	G-P24	4	1
	T-II	1	2	A409	G-TP304, TP304L, TP316, TP316L, TP317, TP321, TP347, TP348	8	1
A376	TP304, TP304H, TP304N, TP316, TP316H, TP316N, TP321, TP321H, TP347, TP347H, TP348, TP348H	8	1				
				A412	T-201, XM-11, XM-19	8	3

Table C1 (Continued)

Base Metal Specification	Grade G- Class C- Type T-	Material No.	Group No.	Base Metal Specification	Grade G- Class C- Type T-	Material No.	Group No.
A414	G-A,B,C,D,E	1	1	A452	G-TP304H, TP316H, TP347H	8	1
	G-F,G	1	2				
A420	G-WPL6	1	1	A455	---	1	2
	G-WPL6	1	2	A473	T-410, 403, 414-T, T-429	6	1
	G-WPL9	9A	1				
	G-WPL3	9B	1		430, 420, 405	6	2
	G-WPL8	11A	1		T-405, 410S, 414-T, 420	7	1
A423	G-1, 2	4	2				
A426	G-CP1, CP2, CP15	3	1		431, T-202, 302's, 303's, 304's, 305, 308, 314, 316's, 317, 321, 347, 348, XM-10	8	1
	G-CP11, CP12	4	1				
	G-CP21, CP22	5	1				
	G-CP5, CP5b, CP7, CP9	5	2				
	G-CPCA15	6	3		XM-11, T-309S, 310, 310S	8	2
A430	G-FP304, FP304H, FP304N, FP316, FP316H, FP316N, FP321, FP321H, FP347, FP347H, FP16-8-2H	8	1	A479	T-403, Cl.1, 410	6	1
					T-414	6	4
					T-405	7	1
					T-XM-8, 430, T-XM-27	7	2
					T-X-M-30 Annealed	7	3
A436	T-1, 1b, 2, 2b, 3, 4, 5, 6	2F			T-302, 304's, 316's, 321's, 347's, 348's	8	1
A439	T-D-2, D-2B, D-2C, D-3, D-3A, D4, D5, D5B	2G			T-309S, T-310S	8	2
					T-XM-11, XM-18, XM-19, XM-29	8	3
A441	---	1	2		G-XM-27	10-I	1
A442	G-55, 60	1	1		UNS 44700	10-J	1
A451	CPF3, CPF3A, CPF3M, G-CPF8, CPF8A, CPF8M, CPF8C, CPF10MC	8	1		UNS 44800	10-K	1
				A487	C-A, AN	1	1
					C-AQ, B, BN, C, CN	1	2
					C-BQ, CQ	1	3
	G-CPH8, PH10, CPK20, CPH20	8	2		C-8N	5	2
					CA15M	6	3

Table C1 (Continued)

Base Metal Specification	Grade G- Class C- Type T-	Material No.	Group No.	Base Metal Specification	Grade G- Class C- Type T-	Material No.	Group No.
A487	G-CA6NM	6	4	A517	G-F	11B	3
	C-1N, 1Q	10A	1		G-B	11B	4
	C-2N, 2Q, 4N	10F	6		G-D	11B	5
	C-4Q, 4QA	11A	3		G-J	11B	6
A500	G-A,B	1	1		G-P	11B	8
	G-C	1	2	A519	G-1008, MT1010, 1012, MT1015, MTX1015, 1016, 1017, 1018, 1019, 1020, MTX1020, 1021, 1022, 1023, 1024, 1025, 1026	1	1
A501	---	1	1				
A508	C-1, C-1a	1	2				
	C-2, 2a, 3, C-3a, 46	3	3				
	C-4, 4A, 5, 5A	11A	5				
A512	G-MT1010, 1011, MT1015, MTX1015, 1016, 1017, 1018, 1020, MT1020, MTX1020, 1025	1	1				
					G-1030, 1035	1	2
					G-4130, 8630	11B	1
	G-1030	1	2	A522	T-I, II	11A	1
A513	G-1008, 1010, MT1010, MT1015, MTX1015, 1016, 1017, 1018, 1019, 1020, MT1020, MTX1020, 1021, 1022, 1023, 1024, 1025, 1026, 1027	1	1	A523	G-A,B	1	1
				A524	G-I, II	1	1
				A526	---	1	1
				A527	---	1	1
				A529	---	1	1
				A533	T-A, C-1, T-B, C-1, T-C, C-1, T-A, C-2, T-B, C-2, T-C, C-2, T-D, C-2, T-D, C-1, T-D, C-2, C-3	3	3
	G-1030, 1033, 1035	1	2				
A514	All Grades	11B	1				
A515	G-55, 60, 65	1	1		T-A, B, C, C-3	11A	4
	G-70	1	2	A537	C-1	1	2
A516	G-55, 60, 65	1	1		C-2	1	3
	G-70	1	2	A539	---	1	1
A517	Grades A,G, H, K, L, Q	11B	1	A542	C-1, 2	11	6
				A543	T-B, C-1, 2, 3	11A	6
	G-E	11B	2		T-C, C-1, 2, 3	11	6

Table C1 (Continued)

Base Metal Specification	Grade G- Class C- Type T-	Material No.	Group No.	Base Metal Specification	Grade G- Class C- Type T-	Material No.	Group No.
A553	T-I, II	11A	1	A660	G-WCB, WCC	1	2
A556	G-A2, B2	1	1	A662	G-A, B	1	1
	G-C2	1	2		G-C	1	2
A557	G-A2, B2	1	1	A663	G-45, 50, 55, 60	1	1
	G-C2	1	2	A669	---	10H	8
A562	---	1	1	A671	G-CA55, CB60, CB65, CC60, CC65, CE55, CE60	1	1
A570	G-30, 33, 36, 40, 45, 50	1	1				
A572	G-42, G-50	1	1		G-CB70, CC70, CD70, CK75	1	2
	60	1	2		G-CD80	1	3
	G-65	1	3	A672	G-A45, A50, A55, B55, B60, B65, C55, C60, C65, E55, E60	1	1
A573	G-58, 65	1	1				
	G-70	1	2		G-B70, C70, D70, N75	1	2
A575	G-1008, 1010, 1012, 1015, 1017, 1020, 1023, 1025	1	1		G-D80	1	3
					G-L65	3	1
A587	---	1	1		G-H75, L70, L75	3	2
A588	All Grades	1	2		G-H80, J80, J90	3	3
A592	G-A	11B	1		G-J100	11A	4
	G-E	11B	2	A675	G-45, 50, 55, 60, 65	1	1
	G-F	11B	3				
A595	G-A, B, C	1	2		G-70	1	2
A606	---	1	2	A688	G-TP304s, TP316s	8	1
A607	G-45, 50, 55, 60, 65, 70	1	2				
A611	G-A, B, C, D, E	1	2		G-TPXM-29	8	3
A612	---	10C	3	A691	G-CMS-75, CMSH-70	1	2
A615	G-40, 60	1	4				
A618	G-I, Ib, II, III	1	2		G-CMSH-80	1	3
A620	---	1	1		G-CM-65, 1/2Cr (55TS)	3	1
A633	G-A, B, C, D	1	2				
	G-E	1	3		G-CM-70, CM-75, 1/2Cr (70TS)	3	2
A645	---	11A	2				
A658	---	10G	1		G-1Cr, 1-1/4Cr	4	1
A660	G-WCA	1	1				

Table C1 (Continued)

Base Metal Specification	Grade G- Class C- Type T-	Material No.	Group No.	Base Metal Specification	Grade G- Class C- Type T-	Material No.	Group No.
A691	G-2-1/4 Cr	5	1	ABS	G-A, B, D, E, DS, CS	1	1
	3Cr, G-5Cr, F5	5	2		G-AH32, DH32, EH32, AH36, DH36, EH36	1	2
A695	T-B, G-35	1	1				
	T-B, G-40, 45	1	2	API 5L	---	1	1
A696	G-B	1	1	API 5LS	G-A, B, X-42, G-X46, X52, X56	1	1
	G-C	1	2				
A706	---	1	4				
A709	G-36	1	1		X60	1	2
	G-50, 50W	1	2		G-X65, X70	1	3
	G-100, 100W	11B	1	API 5LX	G-X42, GX46, X52	1	1
	G-65	1	3				
		3	3		X56, X60	1	2
A724	G-A, B, C	1	4		G-X65, X70	1	3
A727	---	1	1				
A731	G-XM-8, 18Cr-2Mo	7	2				
A737	G-B	1	2	AISI	1005 1006 1008 1009 1010 1011 (SAE) 1012 1013 (SAE) 1015 1016 1017 1018 1019 1020 1021 1022 1023 1025 1026	1	1
	G-C	1	3				
A738	---	1	2				
A739	G-B11	4	1				
	G-B22	5	1				
A765	G-I	1	1				
	G-II	1	2				
	G-III	9B	1				

Table C1 (Continued)

Base Metal Specification	Alloy Designation	Material No.	Base Metal Specification	Alloy Designation	Material No.
B21	C46200, C46400	32	B167	N06600	43
B42	C10200, C12000, C12200	31		N08810	45
			B168	N06600	43
B43	---	32	B169	C61000, C61400	34
B75	C10200, C12000, C12200, C14200	31	B171	C44300, C44400, C44500, C46400, C36500	32
B96	C65500	33		C70600, C71500	34
B98	C65500, C65100	33		C61400, C63000	35
B111	C10200, C12000, C12200, C14200, C19200	31	B209	1060 1100 3003	21
	C44300, C44400, C44500, C28000, C23000, C68700	32		3004 5052 5254 5154 5454 5652	22
	C71500, C70400, C70600	34		6061 Alclad 6061	23
	C60800	35		5083 5086 5456	25
B127	No. 4400	42	B210	1060 3003	21
B133	C10200, C11000, C12000, C12200, C14200	31		5052 5154	22
B135	---	32		6061 6063	23
B148	C95200, C95400	35	B221	1060 1100 3003	21
B150	C61400, C63000, C64200	35		5052 5154 5454	22
B152	C10200, C10400, C10500, C10700, C12200, C12300	31		6061 6063	23
B160	N02200, N02201	41		1060 1100	21
B161	N02200, N02201	41		5154 5454	22
B162	N02200, N02201	41		5083 5456	23
B163	N02200, N02201	41	B234	1060	21
	N04400	42			
	N06600	43			
	N08800, N08810	45			
B164	N04400	42			
B166	N05500	43			

Table C1 (Continued)

Base Metal Specification	Alloy Designation	Material No.	Base Metal Specification	Alloy Designation	Material No.
B234	5052	22	B409	N08800	45
B241	1060	21		N08810	
	1100		B423	N08825	45
	3003		B424	N08825	45
	5052	22	B425	N08825	45
	6061	23	B434	N10003	44
	5083	25	B435	N06002	43
B247	3002	21	B443	N06625	43
	6061	23	B444	N06625	43
	5083	25	B446	N06625	43
B265	G-1, 2	51	B462	UN08020	45
	G-3, 12	52	B463	N08020	45
B308	6061	23		N08825	
B315	---		B464	N080200	45
B333	N10001	44	B466	Nos. 706, 715	34
	N10665		B467	Nos. 706, 715	34
B335	N10001	44	B468	N08020	45
	N10665		B493	R60702	61
B337	G-1, 2, 3	51	B473	N08020	45
	G-3, 12	52	B511	N08330	46
B338	G-1, 2, 7	51	B516	N06600	45
	G-3, 12	52	B517	N06600	45
B348	G-1, 2, 7	51	B523	R60702	61
	G-3, 12	52	B535	N08330	46
B359	Nos. 443, 444, 445, 687	32	B536	N08330	46
	Nos. 704, 706, 710, 715	34	B543	Nos. 704, 706, 715	34
			B550	R60702	61
B395	Nos. 102, 120, 122, 142, 192	31	B551	R60702	61
	Nos. 706, 710, 715	34	B564	N08800	45
B402	C70600	34		N08810	
	C71500		B572	N06002	43
	C72200		B573	N10003	44
B407	N08800	45	B574	N10276	44
B408	N00800	45		N06455	
	N08810		B575	N10276	44
				N06455	

Table C1 (Continued)

Base Metal Specification	Alloy Designation	Material No.	Base Metal Specification	Alloy Designation	Material No.
B581	N06007	45	B621	N08320	45
	N06975		B622	N08320	45
B582	N06007	45	B625	N08904	45
B619	N06002	43	B626	N06007	45
	N06975			N08320	
B620	N08320	45		N06975	

Table C2
Grouping of base metals for qualification

Material No.	Group No.	Std.	Base Metal Specification		Minimum Tensile/Yield, ksi	Type of Base Metal
				Steel and Steel Alloys		
1	1	PS	A31	Grade A	45/23	Rivets (C)
		PS		Grade B	58/29	Rivets (C)
		PS	A36		58/36	Plate (C-Mn-Si)
		PS	A53	Type F	45/25	Pipe
		PS		Type E, Grade A	48/30	Pipe
		PS		Grade B	60/35	Pipe
		PS		Type S, Grade A	48/30	Pipe
		PS		Grade B	60/35	Pipe
		PS	A106	Grade A	48/30	Pipe (C-Si)
		PS		Grade B	60/35	Pipe (C-Si)
		PS	A135	Grade A	48/30	ERW Pipe (C)
		PS		Grade B	60/36	ERW Pipe (C-Mn)
		S	A131	Grade A	58/34	Structural -
		S		Grade B	58/34	Structural -
		S		Grade CS	58/34	Structural -
		S		Grade D	58/34	Structural -
		S		Grade DS	58/34	Structural -
		S		Grade E	58/34	Structural -
		PS	A134			of A-283 and 285
		S	A139	Grade A	48/30	Pipe
		S		Grade B	60/35	Pipe
		S		Grade C	60/42	Pipe
		S		Grade D	60/46	Pipe
		S		Grade E	66/52	Pipe
		P	SA155	Grade C45	45/24	Pipe (C)
		P		Grade C50	50/27	Pipe (C)
		P		Grade C55	55/30	Pipe (C)
		P		Grade KC55	55/30	Pipe (C-Si)
		P		Grade KCF55	55/30	Pipe (C-Si)
		P		Grade KC60	60/32	Pipe (C-Si)
		P		Grade KCF60	60/32	Pipe (C-Si)
		P		Grade KC65	65/35	Pipe (C-Si)
		P		Grade KCF-65	65/35	Pipe (C-Mn-Si)

Table C2 (Continued)

Material No.	Group No.	Std.	Base Metal Specification		Minimum Tensile/Yield, ksi	Type of Base Metal
			Steel and Steel Alloys			
1	1	S	A161	Low Carbon	47/26	Low C Tubes
		S		Grade T1	55/30	C-Mo Tubes
		PS	A178	Grade A	47/26	E-W Tube (C)
				Grade C	60/37	E-W Tube (C)
		PS	A179			Seamless Tubes (Low C)
		PS	A181	Class 60	60/30	Pipe Flanges (C-Si)
				Class 70	70/36	Pipe Flanges (C-Si)
		PS	A192		47/26	Seamless Tubes (C-Si)
		PS	A210	Grade A-1	60/37	Tubes (C)
		PS	A214			ERW Tubes (*C)
		PS	A216	Grade WCA	60/30	Castings (C-Si)
		PS	A226		47/26	E-W Tubes (C-Si)
		PS	A234	WPB	60/35	Pipe Fittings (C-Mn)
		PS	A266	Class 1	60/30	Seamless Forgings (C-Si)
		PS	A283	Grade A	45/25	Plates (C-Si)
				Grade B	50/27	Plates (C-Si)
				Grade C	55/30	Plates (C)
				Grade D	60/33	Plates (C)
		S	A284	Grade C	60/30	Plates (C-Si)
				Grade D	60/33	Plates (C-Si)
		PS	A285	Grade A	45/24	Plates (C)
				Grade B	50/27	Plates (C)
				Grade C	55/30	Plates (C)
		PS	A333	Grade 1	55/30	Pipe (C-Mn)
				Grade 6	60/35	Pipe (C-Mn)
		PS	A334	Grade 1	55/30	Tubes (C-Mn)
				Grade 6	60/35	Tubes (C-Mn-Si)
		PS	A350	Grade LF1	60/30	Forgings (C-Mn)
		PS	A352	Grade LCA	60/30	Castings (C)
				Grade LCB	65/35	Castings (C-Si)
		PS	A369	Grade FPA	48/30	Forgings (C)
				Grade FPB	60/35	Forgings (C)
		PS	A372	Type 1	60/35	Forgings (C)

Table C2 (Continued)

Material No.	Group No.	Std.	Base Metal Specification		Minimum Tensile/Yield, ksi	Type of Base Metal
				Steel and Steel Alloys		
1	1	S	A381	Class Y35	60/35	Pipe
				Class Y46	46/63	Pipe
				Class Y48	48/67	Pipe
				Class Y42	60/42	Pipe
				Class Y50	50/69	Pipe
				Class Y52	52/72	Pipe
				Class Y56	56/75	Pipe
				Class Y60	60/78	Pipe
				Class Y65	65/80	Pipe
		PS	A414	Grade A	45/25	Sheet (C)
				Grade B	50/30	Sheet (C)
				Grade C	55/33	Sheet (C)
				Grade D	60/35	Sheet (C-Mn)
				Grade E	65/38	Sheet (C-Mn)
		PS	A420	Grade WPL6	60/35	Pipe Fittings (C-Mn-Si)
		PS	A442	Grade 55	55/30	Plates (C-Mn-Si)
				Grade 60	60/32	Plates (C-Mn-Si)
		PS	A487	Class A and AN	60/30	Castings (C)
		S	A500	Grade A	45/33	Tubing
				Grade B	58/42	Tubing
		S	A501		58/36	Tubing
		S	A512	Grade MT1010	*	Mechanical Tube
				Grade 1011	*	Mechanical Tube
				Grade MT1015	*	Mechanical Tube
				Grade MTX1015	*	Mechanical Tube
				Grade 1016	*	Mechanical Tube
				Grade 1017	*	Mechanical Tube
				Grade 1018	*	Mechanical Tube
				Grade 1020	*	Mechanical Tube
				Grade MT1020	*	Mechanical Tube
				Grade MTX1020	*	Mechanical Tube
				Grade 1025	*	Mechanical Tube

*Based on condition.

Table C2 (Continued)

Material No.	Group No.	Std.	Base Metal Specification		Minimum Tensile/Yield, ksi	Type of Base Metal
				Steel and Steel Alloys		
1	1	S	A513	Grade 1008	42/30	Mechanical Tube
				Grade 1010	45/32	Mechanical Tube
				Grade MT1015	48/35	Mechanical Tube
				Grade MTX1015	48/35	Mechanical Tube
				Grade 1016		Mechanical Tube
				Grade 1017		Mechanical Tube
				Grade 1018		Mechanical Tube
				Grade 1019		Mechanical Tube
				Grade 1020	52/38	Mechanical Tube
				Grade MT1020	52/38	Mechanical Tube
				Grade MTX1020	52/38	Mechanical Tube
				Grade 1021		Mechanical Tube
				Grade 1022		Mechanical Tube
				Grade 1023		Mechanical Tube
				Grade 1024		Mechanical Tube
				Grade 1025	56/40	Mechanical Tube
				Grade 1026	62/45	Mechanical Tube
				Grade 1027		Mechanical Tube
		PS	A515	Grade 55	55/30	Plates (C-Si)
				Grade 60	60/32	Plates (C-Si)
				Grade 65	65/35	Plates (C-Si)
		PS	A516	Grade 55	55/30	Plates (C-Si)
				Grade 60	60/32	Plates (C-Si)
				Grade 65	65/35	Plates (C-Mn-Si)
		S	A519	Grade 1008	*	Mechanical Tube
				Grade MT1010	*	Mechanical Tube
				Grade 1012	*	Mechanical Tube
				Grade MT1015	*	Mechanical Tube
				Grade MTX1015	*	Mechanical Tube
				Grade 1016		Mechanical Tube
				Grade 1017	*	
				Grade 1018	*	Mechanical Tube
				Grade 1019	*	Mechanical Tube

*Based on condition.

Table C2 (Continued)

Material No.	Group No.	Std.	Base Metal Specification		Minimum Tensile/Yield, ksi	Type of Base Metal
			Steel and Steel Alloys			
I	1	S	A519	Grade 1020	*	Mechanical Tube
				Grade MT1020	*	Mechanical Tube
				Grade 1021	*	Mechanical Tube
				Grade 1022	*	Mechanical Tube
				Grade 1025	*	Mechanical Tube
				Grade 1026	*	Mechanical Tube
		S	A523	Grade A	48/30	Pipe
				Grade B	60/35	Pipe
		PS	A524	Grade I	60/35	Pipe (C-Mn-Si)
				Grade II	55/30	Pipe (C-Mn-Si)
		S	A526			Sheet, Galvanized
		S	A527			Sheet, Galvanized
		S	A529		60/42	Structural
		S	A539		45/35	Coiled Tube
		PS	A556	Grade A2	47/26	Tubes-Seamless (C)
				Grade B2	60/37	Tubes-Seamless (C-Si)
		PS	A557	Grade A2	47/26	Tubes-RW (C)
				Grade B2	60/37	Tubes-RW (C)
		PS	A562		55/30	Plates (C-Cu-Ti)
		S	A570	Grade 30	49/30	Sheet & Strip
				Grade 33	52/33	Sheet & Strip
				Grade 36	53/36	Sheet & Strip
				Grade 40	55/40	Sheet & Strip
				Grade 45	60/45	Sheet & Strip
				Grade 50	65/50	Sheet & Strip
		S	A572	Grade 42	60/42	HSLA, Cb, V Steel
		S	A573	Grade 58	65/35	Plate
				Grade 65	58/32	Plate
			A575	Grade 1008	*	Bars
				Grade 1010	*	Bars
				Grade 1012	*	Bars
				Grade 1015	*	Bars
				Grade 1017	*	Bars

*Based on condition.

Table C2 (Continued)

Material No.	Group No.	Std.	Base Metal Specification		Minimum Tensile/Yield, ksi	Type of Base Metal
			Steel and Steel Alloys			
1	1	S	A75	Grade 1020	*	Bars
				Grade 1023	*	Bars
				Grade i025	*	Bars
		PS	A587		48/30	Pipe (Low-C)
		S	A611	Grade A	42/25	Cold Rolled Sheet
				Grade B	45/30	Cold Rolled Sheet
				Grade C	48/33	Cold Rolled Sheet
				Grade D	52/40	Cold Rolled Sheet
		PS	A620		*	Drawing Quality Sheet
		PS	A660	Grade WCA	60/30	Pipe (C)
		PS	A662	Grade A	58/40	Plate (C-Mn)
				Grade B	65/40	Plate (C-Mn)
		S	A663	Grade 45	45/22.5	Bars (C)
				Grade 50	50/25	Bars (C)
				Grade 55	55/27.5	Bars (C)
				Grade 60	50/30	Bars (C)
				Grade 65		Bars (C)
		PS	A671	Grade CA55	55/30	Pipe (C)
				Grade CE55	55/30	Pipe (C)
				Grade CB60	60/32	Pipe (C-Si)
				Grade CC60	60/32	Pipe (C-Si)
				Grade Ce60	60/32	Pipe (C-Mn-Si)
				Grade CB65	65/35	Pipe (C-Si)
				Grade CC65	65/35	Pipe (C-Mn-Si)
		PS	A672	Grade A45	45/24	Pipe (C)
				Grade A50	50/27	Pipe (C)
				Grade A55	55/30	Pipe (C)
				Grade B55	55/30	Pipe (C-Si)
				Grade C55	55/30	Pipe (C-Si)
				Grade E55	55/30	Pipe (C-Mn-Si)
		PS	A672	Grade B60	60/32	Pipe (C-Si)
				Grade C65	65/35	Pipe (C-Mn-Si)
				Grade C60	60/32	Pipe (C-Si)

*Based on condition.

Table C2 (Continued)

Material No.	Group No.	Std.	Base Metal Specification		Minimum Tensile/Yield, ksi	Type of Base Metal
			Steel and Steel Alloys			
1	1	PS	A672	Grade E60	60/32	Pipe (C-Mn-Si)
				Grade B65	65/35.5	Pipe (C-Si)
		PS	A675	Grade 45	45/22.5	Bars (C)
				Grade 50	50/25	Bars (C)
				Grade 55	55/27.5	Bars (C)
				Grade 60	60/30	Bars (C)
				Grade 65	65/32.5	Bars (C)
		PS	A695	Type B, Grade 35	60/35	Bars (C-Si)
		PS	A696	Grade B	60/35	Bars (C)
		S	A709	Grade 36	58/36	Structural
		PS	A727		60/36	Forging
		PS	A765	Grade I	60/36	Forging
		Q	ABS	Grade A	58/	Structural
				Grade B	58/	Structural
				Grade D	58/	Structural
				Grade E	58/	Structural
				Grade DS	58/	Structural
				Grade CS	58/	Structural
		Q	API 5L	Grade A25	45/	Pipe
				Grade A	48/	Pipe
				Grade B	60/	Pipe
		Q	API 5LS	Grade A	48/	Pipe
				Grade B	60/	Pipe
				Grade X42	60/	Pipe
				Grade X42	60/	Pipe
		Q	API 5LX	Grade X42	60/	Pipe
		Q	1005		Not specified*	Steel Compositions
		Q	1006		Not specified*	Steel Compositions
		Q	1008		Not specified*	Steel Compositions
		Q	1009		Not specified*	Steel Compositions
		Q	1010		Not specified*	Steel Compositions
		Q	1011 (SAE)		Not specified*	Steel Compositions
		Q	1012		Not specified*	Steel Compositions

*Based on condition.

Table C2 (Continued)

Material No.	Group No.	Std.	Base Metal Specification		Minimum Tensile/Yield, ksi	Type of Base Metal
			Steel and Steel Alloys			
1	1	Q	API 5LX	1013 (SAE)	Not specified*	Steel Compositions
		Q		1015	Not specified*	Steel Compositions
		Q		1016	Not specified*	Steel Compositions
		Q		1017	Not specified*	Steel Compositions
		Q		1018	Not specified*	Steel Compositions
		Q		1019	Not specified*	Steel Compositions
		Q		1020	Not specified*	Steel Compositions
		Q		1021	Not specified*	Steel Compositions
		Q		1022	Not specified*	Steel Compositions
		Q		1023	Not specified*	Steel Compositions
		Q		1025	Not specified*	Steel Compositions
		Q		1026	Not specified*	Steel Compositions
1	2	PS	A105		70/36	Pipe Flanges (C-Si)
		PS	A106	Grade C	70/40	Pipe (C-Si)
		S	A131	AH32	68/45.5	Structural
				DH32	68/45.5	Structural
				EH32	68/45.5	Structural
				AH36	70/50	Structural
				DH36	71/50	Structural
				EH36	71/50	Structural
		S	A139	Grade E	66/52	Welded Pipe
		P	A155	Grade KC70	70/38	Pipe (C-Si)
				Grade KCF70	70/38	Pipe (C-Mn-Si)
		PS	A181	Grade 70	70/36	Pipe Flanges (C-Si)
		PS	A210	Grade C	70/40	ERW Tubes (C-Si)
		PS	A216	Grade WCB	70/36	Castings (C-Si)
				Grade WCC	70/40	Castings (C-Mn-Si)
		PS	A234	WPC	70/40	Pipe Fittings (C-Mn)
		S	A242	Type 1	70/50	HSLA Structural
				Type 2	70/50	HSLA Structural
		PS	A266	Class 2	70/36	Seamless Forgings (C-Si)
				Class 4	75/37.5	Seamless Forgings (C-Si)
				Class 3	75/37.5	Seamless Forgings (C-Si)

*Based on condition.

Table C2 (Continued)

Material No.	Group No.	Std.	Base Metal Specification		Minimum Tensile/Yield, ksi	Type of Base Metal
			Steel and Steel Alloys			
1	2	PS	A299		75/42	Plates (C-Mn-Si)
		PS	A350	Grade LF2	70/36	Forgings (C-Mn-Si)
		PS	A352	Grade LCC	70/40	Castings (C-Mn)
		PS	A372	Type II	70/45	Forgings (C)
		S	A381	Class Y46	63/46	Pipe
				Class Y48	67/48	Pipe
				Class Y50	69/50	Pipe
				Class Y52	72/52	Pipe
				Class Y56	75/56	Pipe
				Class Y60	78/60	Pipe
				Class Y65	80/65	Pipe
		PS	A414	Grade F	70/42	Sheet
				Grade G	75/45	Sheet
		PS	A420	Grade WPL6	70/36	Pipe Fittings (C-Mn-Si)
		S	A441		*	HSLA Structural
		PS	A455		70/35	Plates (C-Mn)
		PS	A487	Class AQ	70/30	Castings (C)
				Class B and BN	70/36	Castings (C)
				Class C and CN	70/40	Castings (C)
		S	A500	Grade C	62/46	Structural Tubing
		PS	A508	Class 1	70/36	Forgings (0.35 max. C-Si)
				Class 1A	70/36	Forgings (0.30 max. C-Si)
		S	A512	Grade 1030	**	Mechanical Tube
		S	A513	Grade 1030	**	Mechanical Tube
				Grade 1033	**	Mechanical Tube
				Grade 1035	**	Mechanical Tube
		PS	A515	Grade 70	70/38	Plates (C-Si)
		PS	A516	Grade 70	70/38	Plates (C-Mn-Si)
		S	A519	Grade 1030	**	Mechanical Tube
				Grade 1035	**	Mechanical Tube
		PS	A537	Class 1 (under 2-1/2 in.)	70/50	Plates (C-Mn-Si)
				(over 2-1/2 in.)	65/45	Plates (C-Mn-Si)
		PS	A541	Class 1	70/36	Forgings (C-Si)

*Based on material thickness.
**To be agreed on by manufacturer and purchaser.

Table C2 (Continued)

Material No.	Group No.	Std.	Base Metal Specification		Minimum Tensile/Yield, ksi	Type of Base Metal
			Steel and Steel Alloys			
1	2	PS	A541	Class 1	70/36	Forgings (C-Si)
		PS	A556	Grade C2	70/40	Tubes Seamless (C-Mn)
		S	A557	Grade 62	70/40	RW Tubes (C-Mn)
		S	A572	Grade 50	65/50	HSLA, Cb, V Steel
				Grade 60	75/60	HSLA, Cb, V Steel
		S	A573	Grade 70	70/42	Imp. Tough. Plates
		S	A588	Grade A	70*	HSLA Structural
				Grade B	70*	HSLA Structural
				Grade C	70*	HSLA Structural
				Grade D	70*	HSLA Structural
				Grade E	70*	HSLA Structural
				Grade F	70*	HSLA Structural
				Grade G	70*	HSLA Structural
				Grade H	70*	HSLA Structural
				Grade J	70*	HSLA Structural
				Grade K	70*	HSLA Structural
		S	A595	Grade A	65/55	Tapered Tubes
				Grade B	70/60	Tapered Tubes
				Grade C	70/60	Tapered Tubes
		S	A606		65/45	Sheet and Strip
		S	A607	Grade 45	60/45	Sheet and Strip
				Grade 50	65/50	Sheet and Strip
				Grade 55	70/55	Sheet and Strip
				Grade 60	75/60	Sheet and Strip
				Grade 65	80/65	Sheet and Strip
				Grade 70	85/70	Sheet and Strip
		S	A611	Grade A	42/25	Sheet and Strip
				Grade B	45/30	Sheet and Strip
				Grade C	48/33	Sheet and Strip
				Grade D	52/40	Sheet and Strip
				Grade E	82/80	Sheet and Strip
		S	A618	Grade Ia	70/50	HSLA Tube
				Grade Ib	70/50	HSLA Tube

*Based on material thickness.

Table C2 (Continued)

Material No.	Group No.	Std.	Base Metal Specification		Minimum Tensile/Yield, ksi	Type of Base Metal
				Steel and Steel Alloys		
1	2	S	A618	Grade II	70/50	HSLA Tube
				Grade III	70/50	HSLA Tube
		S	A633	Grade A	63/42	HSLA Steel
				Grade C	65/46	HSLA Steel
				Grade D	65/46	HSLA Steel
		PS	A660	Grade WCB	70/36	Pipe (C)
		PS	A660	Grade WCC	70/40	Pipe (C-Mn-Si)
		S	A662	Grade C	70/43	Plate (C-Mn)
		PS	A671	Grade CB70	70/38	Pipe
				Grade CC70	70/38	Pipe
				Grade CD70	70/46	Pipe
				Grade CK75	75/40	Pipe
			A672	Grade B701	70/38	Pipe
				Grade C70	70/38	Pipe
				Grade D70	70/46	Pipe
				Grade N75	75/40	Pipe
		S	A675	Grade 70	70/35	Bars
				Grade 75	75/37.5	Bars
				Grade 80	80/40	Bars
				Grade 90	90/55	Bars
		PS	A691	Grade CMSH-70	70/50	Pipe
				Grade CMS-75	75/40	Pipe
		S	A695	Grade 40	70/40	Bars
				Grade 45	80/45	Bars
		S	A696	Grade C	70/40	Bars
		S	A709	Grade 50	65/50	Structural
				Grade 50W	70/50	Structural
		PS	A737	Grade B	70/50	Plate
		PS	A738		75/45	Plate
		PS	A765	Grade II	70/36	Forging
		Q	ABS	Grade AH32	68/45	Hull Steel
				Grade DH32	68/45	Hull Steel
				Grade EH32	68/51	Hull Steel
				Grade AH36	71/51	Hull Steel

Table C2 (Continued)

Material No.	Group No.	Std.	Base Metal Specification		Minimum Tensile/Yield, ksi	Type of Base Metal
				Steel and Steel Alloys		
1	2	Q	ABS	Grade DH36	71/51	Hull Steel
				Grade EH36	71/51	Hull Steel
			API 2H		62/42	
		Q	API 5LS	Grade X46	63/46	Pipe
				Grade X52	66/52	Pipe
				Grade X56	71/56	Pipe
				Grade X60	75/60	Pipe
		Q	API 5LX	Grade X46	63/46	Pipe
				Grade X52	66/52	Pipe
				Grade X56	.71/56	Pipe
				Grade X60	75/60	Pipe
1	3	S	A381	Class 60	78/60	Pipe
				Class 65	80/65	Pipe
		PS	A487	Class BQ	80/36	Castings (C)
				Class CQ	80/40	Castings (C)
		PS	A537	Class 2 (under 2-1/2 in.)	80/60	C-Mn-Si Plates
				(over 2-1/2 to 4 in.)	75/55	C-Mn-Si Plates
				(over 4 in. to 6 in.)	70/46	C-Mn-Si Plates
		S	A572	Grade 65	80/65	HSLA, Cb, V Steel
		S	A612		83/*	Plates
		S	A633	Grade E	80/*	HSLA Steel
		S	A663	Grade 70	70/35	Bars
				Grade 75	75/41	Bars
				Grade 80	80/44	Bars
		PS	A671	Grade CD80	80/60	Pipe (C-Mn-Si)
		PS	A672	Grade D80	80/65	Pipe (C-Mn-Si)
		PS	A691	Grade CMSH-80	80/60	Pipe
		PS	A737	Grade C	80/60	Plate
		Q	API 5LS	Grade X70	82/70	Pipe
				Grade X65	77/65	Pipe

*Based on material thickness.

Table C2 (Continued)

Material No.	Group No.	Std.	Base Metal Specification		Minimum Tensile/Yield, ksi	Type of Base Metal
			Steel and Steel Alloys			
1	3	Q	API 5LX	Grade X65	77/65	Pipe
				Grade X70	82/70	Pipe
1	4	S	A615	Grade 40	70/40	Rebar
				Grade 60	90/60	Rebar
		S	A706		80/60	Rebar
2A		S	A48	Class 20	20/*	Gray Iron Castings
				Class 25	25/*	Gray Iron Castings
				Class 30	30/*	Gray Iron Castings
				Class 35	35/*	Gray Iron Castings
				Class 40	40/*	Gray Iron Castings
				Class 60	60/*	Gray Iron Castings
		S	A159	Grade C 1800	18/*	Gray Iron Castings
				Grade G 2500	25/*	Gray Iron Castings
				Grade G 3000	30/*	Gray Iron Castings
				Grade G 3500	35/*	Gray Iron Castings
				Grade G 4000	40/*	Gray Iron Castings
2B		S	A48	Class 45	45/*	Gray Iron Castings
				Class 50	50/*	Gray Iron Castings
				Class 55	55/*	Gray Iron Castings
2C		S	A47	Grade 32510	50/32	Malleable Castings
				Grade 35018	53/35	Malleable Castings
		S	A197		40/30	Malleable Castings
2D		S	A220	Grade 40010	60/40	Malleable Castings
				Grade 45006	65/45	Malleable Castings
				Grade 45008	65/45	Malleable Castings
		S	A602	Grade M3210	50/32	Malleable Castings
				Grade M4504	65/45	Malleable Castings
		S	A220	Grade 50005	70/50	Malleable Castings
2E		S		Grade 60004	80/60	Malleable Castings
		S	A602	Grade M5003	75/50	Malleable Castings
				Grade M5503	75/50	Malleable Castings
2F		S	A436	Type 1	25/*	Gray Iron Castings
				Type 1b	30/*	Gray Iron Castings

*Based on material thickness.

Table C2 (Continued)

Material No.	Group No.	Std.	Base Metal Specification		Minimum Tensile/Yield, ksi	Type of Base Metal
			Steel and Steel Alloys			
2F		S	A436	Type 2	25/*	Gray Iron Castings
				Type 2b	30/*	Gray Iron Castings
				Type 3	25/*	Gray Iron Castings
				Type 4	25/*	Gray Iron Castings
				Type 5	20/*	Gray Iron Castings
				Type 6	25/*	Gray Iron Castings
2G		S	A439	Type D-2	58/30	Ductile Iron Castings
				Type D-2B	58/30	Ductile Iron Castings
				Type D-2C	58/28	Ductile Iron Castings
				Type D-3	55/30	Ductile Iron Castings
				Type D-3A	55/30	Ductile Iron Castings
				Type D-4	60/	Ductile Iron Castings
				Type D-5	55/30	Ductile Iron Castings
				Type D-5B	55/30	Ductile Iron Castings
3	1	P	SA155	Grade 1/2 CR	55/33	Pipe (1/2 Cr-1/2 Mo)
				Grade CM65	65/37	Pipe (C-1/2 Mo)
		PS	A204	Grade A	65/37	Plates (C-1/2 Mo)
		PS	A209	Grade T1	55/30	Tubes (C-1/2 Mo)
				Grade T1a	60/32	Tubes (C-1/2 Mo)
				Grade T1b	53/28	Tubes (C-1/2 Mo)
		PS	A213	Grade T2	60/30	Tubes (1/2 Cr-1/2 Mo)
		PS	A217	Grade WC1	65/35	Castings (C-1/2 Mo)
		PS	A234	Marking WP1	55/30	Pipe Fittings (C-1/2 Mo)
		PS	A242	Type 2	70/50	Plates (C-Mn)
		PS	A250	Grade T1	55/30	Tubes (C-1/2 Mo)
				Grade T1a	60/32	Tubes (C-1/2 Mo)
				Grade T1b	53/28	Tubes (C-1/2 Mo)
		PS	A335	Grade P1	55/30	Pipe (C-1/2 Mo)
				Grade P2	55/30	Pipe (1/2 Cr-1/2 Mo)
				Grade P15	60/30	Pipe (1-1/2 Si-1/2 Mo)
		PS	A352	Grade LC1	65/35	Castings (C-1/2 Mo)
		PS	A369	Grade FP1	55/30	Pipe (C-1/2 Mo)
				Grade FP2	55/30	Pipe (1/2 Cr-1/2 Mo)

*Based on material thickness.

Table C2 (Continued)

Material No.	Group No.	Std.	Base Metal Specification		Minimum Tensile/Yield, ksi	Type of Base Metal
				Steel and Steel Alloys		
3	1	PS	A387	Grade 2, Class 1	55/33	Plate (1/2 Cr-1/2 Mo)
		PS	A426	Grade CP1	55/35	Cast Pipe (C-1/2 Mo)
				Grade CP2	55/30	Cast Pipe (1/2 Cr-1/2 Mo)
				Grade CP15	60/30	Cast Pipe (C-Si-Mo)
		S	A672	Grade L65	65/37	Pipe (C-1/2 Mo)
		PS	A691	Grade CM-65	65/37	Pipe (C-1/2 Mo)
				Grade 1/2 Cr, Class 1	55/33	Pipe (1/2 Cr-1/2 Mo)
3	2	P	A155	Grade CM70	70/40	Pipe (C-1/2 Mo)
				Grade CM75	75/43	Pipe (C-1/2 Mo)
		PS	A182	Grade F1	70/40	Pipe Flanges (C-1/2 Mo)
				Grade F2	70/40	Forgings (1/2 Cr-1/2 Mo)
		PS	A204	Grade B	70/40	Plates (C-1/2 Mo)
				Grade C	75/43	Plates (C-1/2 Mo)
		PS	A234	WP1	70/40	Pipe Fittings (C-1/2 Mo)
		PS	A302	Grade A	75/45	Plates (Mn-Mo)
		PS	A336	Class F1	70/40	Forgings (C-Mo)
		PS	A387	Grade 2, Class 2	70/45	Plate (Cr-Mo)
		PS	A672	Grade L70	70/40	Pipe (C-Mo)
				Grade H75	75/45	Pipe (Mn-Mo)
				Grade L75	75/43	Pipe (C-Mo)
		PS	A691	Grade CM-70	70/40	Pipe (C-Mo)
				Grade CM-75	75/43	Pipe (C-Mo)
				Grade 1/2 Cr (70TS)	55/33	Pipe (1/2 Cr-1/2 Mo)
3	3	PS	A302	Grade B	80/50	Plates (Mn-Mo)
				Grade C	80/50	Plates (Mn-Mo-Ni)
				Grade D	80/50	Plates (Mn-Mo-Ni)
		PS	A508	Class 2	80/50	Forgings (3/4 Ni-Mo-Cr-V)
				Class 2a	90/65	Forgings (1/2 Ni-Cr-Mo)
				Class 3a	90/65	Forgings (1/2 Ni-Cr-Mo)
				Class 3	80/50	Forgings (3/4 Ni-Mo-Cr-V)
				Class 4b	90/70	Forgings (3-1/2 Ni-Cr-Mo-V)
		PS	A533	Type A, Class 1	80/50	Plate (Mn-Mo)
				Type B, Class 1	80/50	Plate (Mn-Mo-Ni)

Table C2 (Continued)

Material No.	Group No.	Std.	Base Metal Specification		Minimum Tensile/Yield, ksi	Type of Base Metal
			Steel and Steel Alloys			
3	3	PS	A533	Type C, Class 1	80/50	Plate (Mn-Mo-Ni)
				Type D, Class 1	80/50	Plate (Mn-Mo-Ni)
				Type A, Class 2	90/70	Plate (Mn-Mo)
				Type B, Class 2	90/70	Plate (Mn-Mo-Ni)
				Type C, Class 2	90/70	Plate (Mn-Mo-Ni)
				Type D, Class 2	90/70	Plate (Mn-Mo-Ni)
		PS	A541	Class 2	80/50	Forgings (3/4 Ni-Mo-Cr-V)
				Class 2A	90/65	Forgings (1/2 Ni-Cr-Mo)
				Class 3	80/50	Forgings (1/2 Ni-Mo-V)
				Class 3A	90/65	Forgings (1/2 Ni-Mo-V)
		PS	A672	Grade H80	80/50	Pipe (Mn-Mo)
				Grade J80	80/50	Pipe (Mn-Mo-Ni)
				Grade J90	90/70	Pipe (Mn-Mo-Ni)
4	1	PS	A155	Grade 1CR	55/33	Pipe (1 Cr-1/2 Mo)
				Grade 1-1/4CR	60/35	Pipe (1 Cr-1/2 Mo-Si)
		PS	A182	Grade F11	70/40	Pipe Flanges (1.25 Cr-1/2 Mo-Si)
				Grade F11a	75/45	Pipe Flanges (1.25 Cr-1/2 Mo-Si)
				Grade F11b	60/30	Pipe Flanges (1.25 Cr-1/2 Mo-Si)
				Grade F12	70/40	Pipe Flanges (1 Cr-1/2 Mo)
				Grade F12b	60/30	Pipe Flanges (1 Cr-1/2 Mo)
		PS	A199	Grade T3b	60/25	Smls. Tubes (2 Cr-1/2 Mo)
				Grade T11	60/25	Smls. Tubes (2 Cr-1/2 Mo)
		PS	A202	Grade A	75/45	Plates (1/2 Cr-1/4 Mo-Si)
				Grade B	85/47	Plates (1/2 Cr-1-1/4 Mn-Si)
		PS	A213	Grade T3b	60/30	Tubes (2 Cr-1/2 Mo)
				Grade T11	60/30	Tubes (1.25 Cr-1/2 Mo-Si)
				Grade T12	60/30	Tubes (2 Cr-1/2 Mo)
		PS	A217	Grade WC4	70/40	Castings (1 Ni-Cr-1/2 Mo)
				Grade WC5	70/40	Castings (3/4 Ni-Mo-3/4 Cr)
				Grade WC6	70/40	Castings (1.25 Cr-1/2 Mo)

Table C2 (Continued)

Material No.	Group No.	Std.	Base Metal Specification		Minimum Tensile/Yield, ksi	Type of Base Metal
				Steel and Steel Alloys		
4	1	PS	A234	WP11	60/30	Pipe Fittings (1.25 Cr-1/2 Mo-Si)
				WP12	60/30	Pipe Fittings (1 Cr-1/2 Mo)
		PS	A333	G-4	60/35	Pipe
		PS	A335	Grade P11	60/30	Pipe (1.25 Cr-1/2 Mo-Si)
				Grade P12	60/30	Pipe (1 Cr-1/2 Mo)
		PS	A336	Class F12	70/40	Forgings (1 Cr-1/2 Mo)
				Class F11	70/40	Forgings (1 Cr-1/2 Mo)
				Class F11A	75/45	Forgings (1 Cr-1/2 Mo)
		PS	A369	Grade FP3b	60/30	Pipes (2 Cr-Mo)
				Grade FP11	60/30	Pipes (1.25 Cr-Mo-Si)
				Grade FP12	60/30	Pipes (1 Cr-Mo)
		PS	A387	Grade 11, Class 1	60/35	Plate (1.25 Cr-1/2 Mo-Si)
				Grade 11, Class 2	75/45	Plate (1.25 Cr-1/2 Mo-Si)
				Grade 12, Class 1	55/33	Plate (1 Cr-1/2 Mo)
				Grade 12, Class 2	65/40	Plate (1 Cr-1/2 Mo)
		S	A389	Grade C23	70/40	Castings
				Grade C24	80/50	Castings
		S	A405	Grade P24	60/30	Pipe
		PS	A426	Grade CP11	70/40	Cast Pipe (1.25 Cr-Mo)
				Grade CP12	60/30	Cast Pipe (2 Cr-Mo)
		PS	A691	Grade 1CF, Class 1	55/33	Pipe (1 Cr-1/2 Mo)
				Grade 1CF, Class 2	70/45	Pipe (1 Cr-1/2 Mo)
				Grade 1-1/4 Cr, Class 1	60/35	Pipe (1-1/4 Cr-1/2 Mo-Si)
				Grade 1-1/4 Cr, Class 2	75/45	Pipe (1-1/4 Cr-1/2 Mo-Si)
		PS	A739	Grade B11	70/45	Bars (1-1/4 Cr-1/2 Mo)
4	2	PS	A333	Grade 4	60/35	Pipes (3/4 Cr-3/4 Ni-Cu-Al)
		PS	A423	Grade 1	60/37	Tubes (3/4 Cr-1/2 Ni-Cu)
				Grade 2	60/37	Tubes (3/4 Ni-1/2 Cu-Mo)
5	1	PS	A155	Grade 2-1/4CR	60/35	Pipe (2.25 Cr-Mo)
		PS	A182	Grade F21	75/45	Pipe Flanges (3 Cr-Mo)
				Grade F22	75/45	Pipe Flanges (2.25 Cr-Mo)
				Grade F22a	60/30	Pipe Flanges (2.25 Cr-Mo)
		PS	A199	Grade T4	60/25	Smls. Tubes (2-1/2 Cr-1/2 Mo-3/4 Si)

Table C2 (Continued)

Material No.	Group No.	Std.	Base Metal Specification		Minimum Tensile/Yield, ksi	Type of Base Metal
				Steel and Steel Alloys		
5	1	PS	A199	Grade T21	60/25	Smls. Tubes (3 Cr-Mo)
				Grade T22	60/25	Smls. Tubes (2.25 Cr-Mo)
		S	A200	Grade T4	60/25	Tubes (Cr-Mo)
				Grade T21	60/25	Tubes (Cr-Mo)
				Grade T22	60/25	Tubes (Cr-Mo)
		PS	A213	Grade T21	60/30	Tubes (3 Cr-Mo)
				Grade T22	60/30	Tubes (2.25 Cr-Mo)
		PS	A217	Grade WC9	70/40	Castings (2.25 Cr-Mo)
		PS	A234	WP22	60/30	Pipe Fittings (2.25 Cr-Mo)
		PS	A335	Grade P21	60/30	Pipe (3 Cr-Mo)
				Grade P22	60/30	Pipe (2.25 Cr-1 Mo)
		PS	A336	Class F21	75/45	Forgings (3 Cr-Mo)
				Class F21a	60/30	Forgings (3 Cr-Mo)
				Class F22	75/45	Forgings (2-1/4 Cr-1 Mo)
				Class F22a	60/30	Forgings (2-1/4 Cr-Mo)
		PS	A369	Grade FP21	60/30	Pipes (3 Cr-Mo)
				Grade FP22	60/30	Pipes (2-1/4 Cr-Mo)
		PS	A387	Grade 21, Class 1	60/30	Plate (3 Cr-Mo)
				Grade 21, Class 2	75/45	Plate (3 Cr-Mo)
				Grade 22, Class 1	60/30	Plate (2-1/4 Cr-Mo)
				Grade 22, Class 2	75/45	Plate (2-1/4 Cr-Mo)
		PS	A426	Grade CP21	60/30	Cast Pipe (3 Cr-Mo)
				Grade CP22	70/40	Cast Pipe (2-1/4 Cr-Mo)
		PS	A691	Grade 2-1/4 Cr, Class 1	60/30	Pipe (2-1/4 Cr-1 Mo)
				Grade 2-1/4 Cr, Class 2	75/45	Pipe (2-1/4 Cr-1 Mo)
				Grade 3 Cr, Class 1	60/30	Pipe (3 Cr-1 Mo)
				Grade 3 Cr, Class 2	75/45	Pipe (3 Cr-1 Mo)
			A739	Grade B22	75/45	Bars (2-1/4 Cr-1 Mo)
5	2	PS	A182	Grade F5	70/40	Pipe Flanges (5 Cr-1/2 Mo)
				Grade F5a	90/65	Pipe Flanges (5 Cr-Mo)
				Grade F7	70/40	Pipe Flanges (7 Cr-Mo)
				Grade F9	85/55	Pipe Flanges (9 Cr-Mo)

Table C2 (Continued)

Material No.	Group No.	Std.	Base Metal Specification		Minimum Tensile/Yield, ksi	Type of Base Metal
				Steel and Steel Alloys		
5	2	PS	A199	Grade T5	60/30	Smls. Tubes (Cr-Mo)
				Grade T7	60/30	Smls. Tubes (7 Cr-Mo)
				Grade T9	60/30	Smls. Tubes (9 Cr-Mo)
		S	A200	Grade T5	60/30	Alloy Tubes
				Grade T7	60/30	Alloy Tubes
				Grade T9	60/30	Alloy Tubes
		PS	A213	Grade T5	60/30	Alloy Tubes (5 Cr-Mo)
				Grade T5b	60/30	Alloy Tubes (5 Cr-Mo-Si)
				Grade T5C	60/30	Alloy Tubes (5 Cr-Mo-Ti)
				Grade T9	60/30	Cr-Mo Tubes (9 Cr-Mo)
				Grade T7	60/30	Cr-Mo Tubes (7 Cr-Mo)
		PS	A217	Grade C5	90/60	Castings (5 Cr-Mo)
				Grade C12	90/60	Castings (9 Cr-Mo)
		PS	A234	Grade WP5	60/30	Pipe Fittings (5 Cr-1/2 Mo)
				Grade WP7	60/30	Pipe Fittings (7 Cr-1/2 Mo)
				Grade WP9	60/30	Pipe Fittings (9 Cr-1/2 Mo)
		PS	A335	Grade P5	60/30	Pipe (5 Cr-1/2 Mo)
				Grade P5b	60/30	Pipe (5 Cr-1/2 Mo-Si)
				Grade P5c	60/30	Pipe (5 Cr-1/2 Mo-Ti)
				Grade P9	60/30	Pipe (9 Cr-1/2 Mo)
				Grade P7	60/30	Pipe (7 Cr-1/2 Mo)
		PS	A336	Class F5	60/36	Forgings (5 Cr-1/2 Mo)
				Class F5a	80/56	Forgings (5 Cr-1/2 Mo)
				Class F9	85/55	Forgings (9 Cr-1 Mo-3/4 Si)
		PS	A369	Grade FP5	60/30	Forgings (5 Cr-1/2 Mo)
				Grade FP7	60/30	Pipes (7 Cr-1/2 Mo)
				Grade FP9	60/30	Pipes (9 Cr-1/2 Mo)
		PS	A387	Grade 5, Class 1	60/30	Plate (5 Cr-1/2 Mo)
				Grade 5, Class 2	75/45	Plate (5 Cr-1/2 Mo)
		PS	A426	Grade CP5	90/60	Cast Pipe (5 Cr-1/2 Mo)
				Grade CP5b	60/30	Cast Pipe (5 Cr-1.5 Si-1/2 Mo)
				Grade CP7	60/30	Cast Pipe (7 Cr-1/2 Mo)
				Grade CP9	90/60	Cast Pipe (9 Cr-1 Mo)
				Grade CP22	70/40	Cast Pipe (2-1/2 Cr-1 Mo)

Table C2 (Continued)

Material No.	Group No.	Std.	Base Metal Specification		Minimum Tensile/Yield, ksi	Type of Base Metal
			Steel and Steel Alloys			
5	2	PS	A487	Grade 8N	85/55	Castings (2-1/4 Cr-Mo)
5	2	S	A542	Class 3	99/75	Plate (2-1/4 Cr-Mo)
				Class 4	85/60	Plate (2-1/4 Cr-Mo)
		PS	A691	Grade 5 Cr, Class 1	60/30	Pipe (5 Cr-1/2 Mo)
				Grade 5 Cr, Class 2	75/45	Pipe (5 Cr-1/2 Mo)
				Grade F5	70/40	Flanges (5 Cr-1/2 Mo)
5	3	S	A542	Class 1	105/85	Plate (2-1/4 Cr-Mo)
				Class 2	115/100	Plate (2-1/4 Cr-Mo)
6	1	PS	A182	Grade F6A, Class 1	70/40	Flanges (13 Cr)
				Grade F6A, Class 4	130/110	Flanges (13 Cr)
				Grade F6B	110/90	Flanges (13 Cr-5 Mo)
				Grade F6MN	110/90	Flanges (13 Cr-4 Ni)
		PS	A240	Type 410	65/30	Plate (13 Cr)
		PS	A268	Grade TP410	60/30	Tubes (13 Cr)
				Grade TP409	60/30	Tubes (11 Cr-Ti)
		PS	A479	Type 403, Class 1	70/40	Bars and Shapes
				Type 410, Class 1	70/40	Bars and Shapes
		S	A473	Type 410	70/40	Forgings
				Type 403	70/40	Forgings
				Type 414-T	115/90	Forgings
6	2	PS	A182	Grade F429	60/35	Flanges (15 Cr)
		PS	A240	Type 429	65/30	Plate (15 Cr)
		PS	A268	Grade TP429	60/35	Tubes (15 Cr)
		S	A473	Type 429	65/35	Forgings
				Type 430	70/35	Forgings
				Type 420	70/35	Forgings
				Type 405	60/30	Forgings
6	3	PS	A182	Grade F6a, Class 2	85/55	Forgings (13 Cr)
				Grade F6a, Class 3	110/185	Forgings (13 Cr)
				Grade F6b	110/190	Forgings (13 Cr-1/2 Mo)
		PS	A217	Grade CA-15	90/65	Castings (13 Cr-Mo-C max)
		PS	A336	Glass F6	85/55	Forgings (13 Cr)
		PS	A426	Grade CPCA15	90/65	Cast Pipe (13 Cr-Mo)
		PS	A479	Type 414, tempered	115/90	Bars and Shapes
		PS	A487	Class CA15M	90/65	Castings (13 Cr)
6	4	PS	A487	Grade CA6NM	110/80	Castings (13 Cr-4 Ni)
		PS	A182	Grade F6NM	110/90	Flanges (13 Cr-4 Ni)

Table C2 (Continued)

Material No.	Group No.	Std.	Base Metal Specification		Minimum Tensile/Yield, ksi	Type of Base Metal
				Steel and Steel Alloys		
7	1	PS	A240	Type 405	60/25	Plate (12 Cr-Al)
				Type 409	55/30	Plate (11 Cr)
				Type 410S	60/30	Plate (13 Cr)
		PS	A268	Grade TP405	60/30	Tubes (12 Cr-Al)
				Grade TP409	60/30	Plate (11 Cr-Ti)
		S	A473	Type 405	60/30	Forgings
				Type 410S	65/35	Forgings
				Type 414, tempered	115/90	Forgings
				Type 420	*	Forgings
				Type 431, tempered	115/90	Forgings
		S	A479	Type 405	60/25	Bars and Shapes
7	2	PS	A182	Grade F430	60/35	Pipe Flanges (17 Cr)
		PS	A240	Type XM-8	65/30	Plate (18 Cr-Ti)
				S44400	60/40	Plate (18 Cr-2 Mo-Ti)
				Type 430	65/30	Plate (17 Cr)
				Type XM-27	65/40	Plate (26 Cr-1/2 Mo)
				Type XM-33	68/45	Plate (26 Cr-1-1/4 Mo-1/2 Ti)
		PS	A268	Grade TP430	60/35	Tubes (17 Cr)
				Grade TPXM-8	60/30	Tubes (18 Cr-Ti)
				Type 18 Cr-2M	60/45	Tubes (18 Cr-2 Mo-Ti)
		PS	A479	Type XM-8	70/40	Bars and Shapes
				Type XM-27	65/40	Bars and Shapes (1 Mo)
				Type 430	70/40	Bars and Shapes (17 Cr)
		PS	A731	Type (18Cr-2Mo)	60/40	Smls. and Pipe
				TPXM-8	60/30	Smls. and Pipe
7	3	PS	A479	Type XM-30	70/40	Bars and Shapes
8	1	S	A167	Type 301	75/30	Plate
				Type 302	75/30	Plate
				Type 302B	75/30	Plate
				Type 304	75/30	Plate
				Type 304L	70/25	Plate
				Type 304LN	75/30	Plate
				Type 305	70/25	Plate

*Fully hardened to Rockwell Hardness C50.

Table C2 (Continued)

Material No.	Group No.	Std.	Base Metal Specification		Minimum Tensile/Yield, ksi	Type of Base Metal
			Steel and Steel Alloys			
8	1	S	A167	Type 308	75/30	Plate
				Type 309	75/30	Plate
				Type 309S	75/30	Plate
				Type 309Cb	75/30	Plate
				Type 310	75/30	Plate
				Type 310S	75/30	Plate
				Type 310Cb	75/30	Plate
		S	A167	Type 316	75/30	Plate
				Type 316LN	75/30	Plate
				Type 316L	70/25	Plate
				Type 316Cb	75/30	Plate
				Type 317	75/30	Plate
				Type 317L	75/30	Plate
				Type 321	75/30	Plate
				Type 321H	75/30	Plate
				Type 347	75/30	Plate
				Type 348	75/30	Plate
				Type XM-15	75/30	Plate
		PS	A182	Grade F304	75/30	Pipe Flanges (Cr-Ni)
				Grade F304H	75/30	Pipe Flanges (Cr-Ni)
				Grade F304L	70/25	Pipe Flanges (Cr-Ni)
				Grade F304N	80/35	Pipe Flanges (Cr-Ni-N)
				Grade F316	75/30	Pipe Flanges (Cr-Ni-Mo)
				Grade F316H	75/30	Pipe Flanges (Cr-Ni-Mo)
				Grade F316L	65/25	Pipe Flanges (Cr-Ni-Mo)
				Grade F316N	80/35	Pipe Flanges (Cr-Ni-Mo-N)
				Grade F321	75/30	Pipe Flanges (Cr-Ni-Ti)
				Grade F321H	75/30	Pipe Flanges (Cr-Ni-Ti)
				Grade F347	75/30	Pipe Flanges (Cr-Ni-Cb)
				Grade F347H	75/30	Pipe Flanges (Cr-Ni-Cb)
				Grade F348	75/30	Pipe Flanges (Cr-Ni-Cb)
				Grade F348H	75/30	Pipe Flanges (Cr-Ni-Cb)
		PS	A213	Grade TP304	75/30	AISI Smls. Tubes (Cr-Ni)
				Grade TP304H	75/30	AISI Smls. Tubes (Cr-Ni)

Table C2 (Continued)

Material No.	Group No.	Std.	Base Metal Specification		Minimum Tensile/Yield, ksi	Type of Base Metal
				Steel and Steel Alloys		
8	1	PS	A213	Grade TP304L	70/24	AISI Smls. Tubes (Cr-Ni)
				Grade TP304N	80/35	Alloy Smls. (Cr-Ni-N)
				Grade TP316	75/30	AISI Smls. Tubes (Cr-Ni-Mo)
				Grade TP316H	75/30	AISI Smls. Tubes (Cr-Ni-Mo)
				Grade TP316L	70/25	AISI Smls. Tubes (Cr-Ni-Mo)
				Grade TP316N	80/35	Tubes (Cr-Ni-Mo-N)
				Grade TP321	75/30	AISI Smls. Tubes (Cr-Ni-Ti)
				Grade TP321H	75/30	AISI Smls. Tubes (Cr-Ni-Ti)
				Grade TP347	75/30	AISI Smls. Tubes (Cr-Ni-Cb)
				Grade TP347H	75/30	AISI Smls. Tubes (Cr-Ni-Cb)
				Grade TP348	75/30	AISI Smls. Tubes (Cr-Ni-Cb)
				Grade TP348H	75/30	AISI Smls. Tubes (Cr-Ni-Cb)
				Grade XM-15	75/30	Smls. Tubes (Cr-Ni-Si)
		PS	A240	Type 302	75/30	Plate (Cr-Ni)
				Type 304	75/30	AISI Plate (Cr-Ni)
				Type 304H	75/30	AISI Plate (Cr-Ni)
				Type 304L	70/25	AISI Plate (Cr-Ni)
				Type 304N	80/35	AISI Plate (Cr-Ni-N)
				Type 316	75/30	AISI Plate (Cr-Ni-Mo)
				Type 316 H	75/30	AISI Plate (Cr-Ni-Mo)
				Type 316L	70/25	AISI Plate (Cr-Ni-Mo)
				Type 316N	80/35	AISI Plate (Cr-Ni-Mo-N)
				Type 316Cb	75/30	AISI Plate (Cr-Ni-Mo)
				Type 316Ti	75/30	AISI Plate (Cr-Ni-Mo)
				Type 317	75/30	AISI Plate (Cr-Ni-Mo)
				Type 317L	75/30	AISI Plate (Cr-Ni-Mo)
				Type 321	75/30	AISI Plate (Cr-Ni-Ti)
				Type 321H	75/30	AISI Plate (Cr-Ni-Ti)
				Type 347	75/30	AISI Plate (Cr-Ni-Cb)
				Type 347H	75/30	AISI Plate (Cr-Ni-Cb)
				Type 348	75/30	AISI Plate (Cr-Ni-Cb)
				Type 348H	75/30	AISI Plate (Cr-Ni-Cb)
				Type XM-15	75/30	Alloy Plate (Cr-Ni-Si)
				Type XM-21	85/40 90/50	Plate (Cr-Ni-N) Sheet and Strip

Table C2 (Continued)

Material No.	Group No.	Std.	Base Metal Specification		Minimum Tensile/Yield, ksi	Type of Base Metal
			Steel and Steel Alloys			
8	1	PS	A249	Grade TP304	75/30	AISI Welded Tubes (Cr-Ni)
				Grade TP304H	75/30	AISI Welded Tubes (Cr-Ni)
				Grade TP304L	70/25	AISI Welded Tubes (Cr-Ni)
				Grade TP304N	80/35	Welded Tubes (Cr-Ni-N)
				Grade TP316	75/30	AISI-316 Welded Alloy Steel Tubes (16 Cr-12 Ni-2 Mo)
				Grade TP316N	75/30	AISI Welded Tubes (Cr-Ni-Mo-N)
				Grade TP316L	70/25	AISI Welded Tubes (Cr-Ni-Mo)
				Grade TP316H	80/35	AISI Welded Tubes (Cr-Ni-Mo)
				Grade TP317	75/30	AISI Welded Tubes (Cr-Ni-Mo)
				Grade TP321	75/30	AISI Welded Tubes (Cr-Ni-Ti)
				Grade TP321H	75/30	AISI Welded Tubes (Cr-Ni-Ti)
				Grade TP347	75/30	AISI Welded Tubes (Cr-Ni-Cb)
				Grade TP347H	75/30	AISI Welded Tubes (Cr-Ni-Cb)
				Grade TP348	75/30	AISI Welded Tubes (Cr-Ni-Cb)
				Grade TP348H	75/30	AISI Welded Tubes (Cr-Ni-Cb)
				Grade TPXM-15	75/30	Welded Tubes (Cr-Ni-Si)
		S	A269	Grade TP304	*	Tubing
				Grade TP304L	*	Tubing
				Grade TP304LN	*	Tubing
				Grade TP316	*	Tubing
				Grade TP316L	*	Tubing
				Grade TP316LN	*	Tubing
				Grade TP317	*	Tubing
				Grade TP321	*	Tubing
				Grade TP347	*	Tubing

*Tensile-yield strength not specified in A269.

Table C2 (Continued)

Material No.	Group No.	Std.	Base Metal Specification		Minimum Tensile/Yield, ksi	Type of Base Metal
			Steel and Steel Alloys			
8	1	S	A269	Grade TP348	*	Tubing
				Grade TPXM-10	*	Tubing
				Grade TPXM-11	*	Tubing
				Grade TPXM-15	*	Tubing
				Grade TPXM-19	*	Tubing
				Grade TPXM-29	*	Tubing
		S	A270	Type 304	75/30	Tubing
		S	A271	Grade TP304	75/30	Tubing
				Grade TP304H	75/30	Tubing
				Grade TP316	75/30	Tubing
				Grade TP316H	75/30	Tubing
				Grade TP321	75/30	Tubing
				Grade TP321H	75/30	Tubing
				Grade TP347	75/30	Tubing
		S	A271	Grade TP347H	75/30	Tubing
		PS	A312	Grade TP304	75/30	Pipe (Cr-Ni)
				Grade TP304H	75/30	Pipe (Cr-Ni)
				Grade TP304L	70/25	Pipe (Cr-Ni)
				Grade TP304N	80/35	Smls. and Welded Pipe (Cr-Ni-N)
				Grade TP316	75/30	Pipe (16 Cr-12 Ni-2 Mo)
				Grade TP316H	75/30	Pipe (16 Cr-12 Ni-2 Mo)
				Grade TP316L	70/25	Pipe (16 Cr-12 Ni-2 Mo)
				Grade TP316N	80/25	Smls. and Welded Pipe (Cr-Ni-Mo-N)
				Grade TP317	75/30	Pipe (Cr-Ni-Mo)
				Grade TP321	75/30	Pipe (Cr-Ni-Ti)
				Grade TP321H	75/30	Pipe (Cr-Ni-Ti)
				Grade TP347	75/30	Pipe (Cr-Ni-Cb)
				Grade TP347H	75/30	Pipe (Cr-Ni-Cb)
				Grade TP348	75/30	Pipe (Cr-Ni-Cb)
				Grade TP348H	75/30	Pipe (Cr-Ni-Cb)
				Grade TPXM-15	75/30	Pipe (Cr-Ni)

*Tensile-yield strength not specified in A269.

Table C2 (Continued)

Material No.	Group No.	Std.	Base Metal Specification		Minimum Tensile/Yield, ksi	Type of Base Metal
			Steel and Steel Alloys			
8	1	PS	A336	Class F304	70/30	Forgings (Cr-Ni)
				Class F304H	70/30	Forgings (Cr-Ni)
				Class F304L	65/25	Forgings (Cr-Ni)
				Class 304N	80/35	Forgings (Cr-Ni)
				Class F316	70/30	Forgings (Cr-Ni-Mo)
				Class F316H	70/30	Forgings (Cr-Ni-Mo)
				Class F316L	65/25	Forgings (Cr-Ni-Mo)
				Class F316N	80/35	Forgings (Cr-Ni-Mo)
				Class F321	70/30	Forgings (Cr-Ni-Ti)
				Class F321H	70/30	Forgings (Cr-Ni-Ti)
				Class F347	70/30	Forgings (Cr-Ni-Cb)
				Class F347H	70/30	Forgings (Cr-Ni-Cb)
				Class F348	70/30	Forgings (Cr-Ni-Cb)
				Class 348H	65/25	Forgings (Cr-Ni-Cb)
				Class FXM-11	90/50	Forgings (Cr-Ni-Mn-N)
				Class FXM-19	100/90	Forgings (Cr-Ni-Mo-Mn-Cb-N)
		PS	A351	Grade CF3	70/30	Castings (Cr-Ni)
				Grade CF3A	77/35	Castings (Cr-Ni)
				Grade CF8	70/30	Castings (Cr-Ni)
				Grade CF8A	77/35	Castings (Cr-Ni)
				Grade CF3M	70/30	Castings (Cr-Ni-Mo)
				Grade CF8M	70/30	Castings (Cr-Ni-Mo)
				Grade CF8C	70/30	Castings (Cr-Ni-Cb)
				Grade CF10	70/30	Castings (Cr-Ni-Mo)
				Grade CF10M	70/30	Castings (Cr-Ni-Mo)
				Grade CE8M	75/35	Castings (Cr-Ni-Mo)
		PS	A358	Grade 304	75/30	AISI Pipe (Cr-Ni)
				Grade 304H	70/30	AISI Pipe (Cr-Ni)
				Grade 304L	65/25	AISI Pipe (Cr-Ni)
				Grade 304N	80/35	AISI Pipe (Cr-Ni-N)
				Grade 316	75/30	AISI Pipe (Cr-Ni-Mo)
				Grade 316H	70/30	AISI Pipe (Cr-Ni-Mo)
				Grade 316L	65/25	AISI Pipe (Cr-Ni-Mo)

Table C2 (Continued)

Material No.	Group No.	Std.	Base Metal Specification		Minimum Tensile/Yield, ksi	Type of Base Metal
			Steel and Steel Alloys			
8	1	PS	A358	Grade 316N	80/35	AISI Pipe (Cr-Ni-Mo-Ni)
				Grade 321	75/30	AISI Pipe (Cr-Ni-Ti)
				Grade 347	75/30	AISI Pipe (Cr-Ni-Cb)
				Grade 348	75/30	AISI Pipe (Cr-Ni-Cb)
		PS	A376	Grade TP304	75/30	AISI Smls. Tubes (Cr-Ni)
				Grade TP304H	75/30	AISI Smls. Tubes (Cr-Ni)
				Grade TP304N	80/35	Smls. Pipe (Cr-Ni-N)
				Grade TP316	75/30	AISI Smls. Tubes (Cr-Ni-Mo)
				Grade TP316H	75/30	AISI Smls. Tubes (Cr-Ni-Mo)
				Grade TP316N	80/35	Smls. Pipe (Cr-Ni-Mo-N)
				Grade TP321	75/30	AISI Smls. Tubes (Cr-Ni-Ti)
				Grade TP321H	75/30	AISI Smls. Tubes (Cr-Ni-Ti)
				Grade TP347	75/30	AISI Smls. Tubes (Cr-Ni-Cb)
				Grade TP347H	75/30	AISI Smls. Tubes (Cr-Ni-Cb)
				Grade TP348	75/30	AISI Smls. Tubes (Cr-Ni-Cb)
				Grade TP348H	75/30	AISI Smls. Tubes (Cr-Ni-Cb)
		PS	A403	WP304	75/30	Pipe Fittings
				WP304H	75/30	Pipe Fittings
				WP304HF	75/30	Pipe Fittings
				WP304L	70/30	Pipe Fittings
				WP304N	80/35	Pipe Fittings
				WP316	75/30	Wrought Pipe Fittings
				WP316H	75/30	Wrought Pipe Fittings
				WP316HF	75/30	Pipe Fittings
				WP316L	70/25	Pipe Fittings
				WP316N	80/35	Pipe Fittings
				WP317	75/30	Pipe Fittings
				WP321	75/30	Pipe Fittings
				WP321H	75/30	Pipe Fittings
				WP321HF	75/30	Pipe Fittings
				WP347	75/30	Pipe Fittings
				WP347H	75/30	Pipe Fittings
				WP347HF	75/30	Pipe Fittings

Table C2 (Continued)

Material No.	Group No.	Std.	Base Metal Specification		Minimum Tensile/Yield, ksi	Type of Base Metal
				Steel and Steel Alloys		
8	1	PS	A403	WP348	75/30	Pipe Fittings
				WP348H	75/30	Pipe Fittings
		PS	A409	TP304	75/30	Welded Pipe
				TP304L	70/25	Welded Pipe
				TP316	75/30	Welded Pipe
				TP316L	70/25	Welded Pipe
				TP317	75/30	Welded Pipe
				TP321	75/30	Welded Pipe
				TP347	75/30	Welded Pipe
				TP348	75/30	Welded Pipe
		PS	A430	Grade FP304	70/30	Pipe (Cr-Ni)
				Grade FP304H	70/30	Pipe (Cr-Ni)
				Grade FP304N	75/35	Smls. Pipe (Cr-Ni-N)
				Grade FP316H	70/30	Pipe (Cr-Ni-Mo)
				Grade FP316N	75/35	Pipe (Cr-Ni-Mo)
				Grade FP321	75/30	Pipe (Cr-Ni-Ti)
				Grade FP321H	70/30	Pipe (Cr-Ni-Ti)
				Grade FP347	70/30	Pipe (Cr-Ni-Cb)
				Grade FP347H	70/30	Pipe (Cr-Ni-Cb)
				Grade FP16-82H	70/30	Pipe (16 Cr-8 Ni-2 Mo)
		PS	A451	Grade CPF3	70/30	Castings (Cr-Ni)
				Grade CPF3H	77/35	Castings (Cr-Ni)
				Grade CPF3M	70/30	Castings (Cr-Ni)
				Grade CPF8	70/30	Castings (Cr-Ni)
				Grade CPF8A	77/35	Castings (Cr-Ni, ferrite)
				Grade CPF8M	70/30	Castings (Cr-Ni-Mo)
				Grade CPF8C	70/30	Castings (Cr-Ni-Cb)
				Grade CPF10MC	70/30	Castings (Cr-Ni-Mo-Cb)
		PS	A452	Grade TP304H	75/30	Cast Pipe (Cr-Ni)
				Grade TP316H	75/30	Cast Pipe (Cr-Ni-Mo)
				Grade TP347H	75/30	Cast Wrought Pipe (Cr-Ni-Cb)

Table C2 (Continued)

Material No.	Group No.	Std.	Base Metal Specification		Minimum Tensile/Yield, ksi	Type of Base Metal
				Steel and Steel Alloys		
8	1	S	A473	Type 202	90/45	Forgings
				Type 302	75/30	Forgings
				Type 302B	75/30	Forgings
				Type 303	75/30	Forgings
				Type 303Se	75/30	Forgings
				Type 304	75/30	Forgings
				Type 304L	65/25	Forgings
				Type 305	75/30	Forgings
				Type 308	75/30	Forgings
				Type 314	75/30	Forgings
				Type 316	75/30	Forgings
				Type 316L	65/25	Forgings
				Type 317	75/30	Forgings
				Type 321	75/30	Forgings
				Type 347	75/30	Forgings
				Type 348	75/30	Forgings
				Type XM-10	90/50	Forgings
				Type XM-11	90/50	Forgings
		PS	A479	Type 302	75/30	Bars and Shapes (Cr-Ni)
				Type 304	75/30	Bars and Shapes (Cr-Ni)
				Type 304H	75/30	Bars and Shapes
				Type 304L	70/30	Bars and Shapes (Cr-Ni)
				Type 304N	80/30	Bars and Shapes
				Type 316	75/30	Bars and Shapes (Cr-Ni-Mo)
				Type 316H	75/30	Bars and Shapes
				Type 316L	70/25	Bars and Shapes (Cr-Ni-Mo)
				Type 316N	80/35	Bars and Shapes
				Type 321	75/30	Bars and Shapes (Cr-Ni-Ti)
				Type 321H	75/30	Bars and Shapes
				Type 347	75/30	Bars and Shapes (Cr-Ni-Cb)
				Type 347H	75/30	Bars and Shapes
				Type 348	75/30	Bars and Shapes (Cr-Ni-Cb)
				Type 348H	75/30	Bars and Shapes

Table C2 (Continued)

Material No.	Group No.	Std.	Base Metal Specification		Minimum Tensile/Yield, ksi	Type of Base Metal
				Steel and Steel Alloys		
8	1	PS	A688	Grade TP304	75/30	Tubes-Welded (Cr-Ni)
				Grade TP304L	70/25	Tubes-Welded (Cr-Ni)
				Grade TP316	75/30	Tubes-Welded (Cr-Ni-Mo)
				Grade TP316L	70/25	Tubes-Welded (Cr-Ni-Mo)
8	2	PS	A182	Grade F310	75/30	Pipe Flanges (Cr-Ni)
				Grade F10	80/30	Pipe Flanges (Cr-Ni)
				Grade F45	87/45	Pipe Flanges (Cr-Ni-Si-N)
		PS	A213	Grade TP310	75/30	AISI Smls. Tubes (Cr-Ni)
		PS	A240	Type 309s	75/30	AISI Plate (Cr-Ni)
				Type 309Cb	75/30	Plate (Cr-Ni-Cb)
				Type 310s	75/30	Plate (Cr-Ni)
				Type 310Cb	75/30	Plate (Cr-Ni-Cb)
		PS	A249	Grade TP309	75/30	Pipe (Cr-Ni)
				Grade TP310	75/30	Tubes (Cr-Ni)
		PS	A312	Grade TP309	75/30	Pipe (Cr-Ni)
				Grade TP310	75/30	Pipe (Cr-Ni)
		PS	A336	Class F310	75/30	Forgings (Cr-Ni)
				Class FXM-11	90/50	Forgings (Cr-Ni)
				Class FXM-19	100/55	Forgings (Cr-Ni)
		PS	A351	Grade CH8	65/28	Castings (Cr-Ni)
				Grade CH20	70/30	Castings (Cr-Ni)
				Grade CK20	65/28	Castings (Cr-Ni)
				Grade CN7M	62/25	Castings (Ni-Cr-Cu-Mo)
		PS	A358	Grade 309	75/30	AISI Pipe (Cr-Ni)
				Grade 310	75/30	Pipe (Cr-Ni)
		PS	A403	WP309	75/30	Pipe Fittings (Cr-Ni)
				WP310	75/30	Pipe Fittings (Cr-Ni)
		PS	A409	Grade TP309	75/30	Pipe (Cr-Ni)
				Grade TP310	75/30	Pipe (Cr-Ni)
		PS	A451	Grade CPH8	65/28	Castings (Cr-Ni)
				Grade CPH10	70/30	Castings (Cr-Ni)
				Grade CPK20	65/28	Castings (Cr-Ni)
				Grade CPH20	70/30	Castings (Cr-Ni)

Table C2 (Continued)

Material No.	Group No.	Std.	Base Metal Specification		Minimum Tensile/Yield, ksi	Type of Base Metal
				Steel and Steel Alloys		
8	2	S	A473	Type 309	75/30	Forgings
				Type 309s	75/30	Forgings
				Type 310	75/30	Forgings
				Type 310s	75/30	Forgings
		PS	A479	Type 310S	75/30	Bars and Shapes (Cr-Ni)
				Type 309S	75/30	Bars and Shapes (Cr-Ni)
8	3	PS	A182	Grade FXM-19	100/55	Forgings (Cr-Ni-Mn-Mo-Cb-N-V)
				Grade FXM-11	90/50	(20 Cr-6 Ni-9 Mo)
				Grade F44	94/44	(20 Cr-8 Ni-6 Mo-Low C)
				Grade FR	63/46	(2 Ni-1 Cu)
		PS	A240	Type XM-17	90/50	Plate (Cr-Ni-Mn-Mo-N)
					100/60	Sheet and Strip (Cr-Ni-Mn-Mo)
				Type XM-18	90/50	Plate (Low C-Cr-Ni-Mn-Mo)
					100/60	Sheet and Strip (Low C-Cr-Ni-Mn-Mo)
				Type XM-19	100/55	Plate (Cr-Ni-Mn-N-Cb-V)
					120/75	Sheet and Strip (Cr-Ni-Mn-N-Cb-V)
				Type XM-29	100/55	Plate (Cr-Ni-Mn-N)
		PS	A249	Grade TPXM-19	100/55	Welded Tubes (Cr-Ni-Mn-Mo-Cb-N-V)
				Grade TPXM-29	100/55	Tube (Cr-Ni-Mn-N)
		PS	A312	Grade TPXM-19	100/55	Pipe
				Type XM-29	100/55	Pipe (Cr-Ni-Mn-N)
				Type XM-11	90/50	Pipe (Cr-Ni-Mn-N)
		PS	A336	Class XM-11	90/50	Forgings (21 Cr-6 Ni-9 Mn)
				Class FXM-19	100/55	Forgings (22 Cr-13 Ni-9 Mn)
			A351	Grade CG6MMN	75/35	Castings (22 Cr-12 Ni-5 Mn-N-Cb-V)
			A358	Grade XM-19	100/55	Pipe (22 Cr-13 Ni-5 Mn)
				Grade XM-29	100/55	Pipe (18 Cr-3 Ni-12 Mn)
			A403	Marking WPXM-19	100/55	Pipe Fittings (22 Cr-12 Ni-5 Mn-2 Mo-Cb-N-V)

Table C2 (Continued)

Material No.	Group No.	Std.	Base Metal Specification		Minimum Tensile/Yield, ksi	Type of Base Metal
			Steel and Steel Alloys			
8	3	PS	A412	Type 201	95/45	Plate, Sheet Strip
				Type XM-11	90/50	Plate, Sheet Strip
				Type XM-19	100/55	Plate
					120/75	Sheet and Strip
		PS	A479	Type XM-11	90/50	Bars and Shapes
		PS	A479	Type XM-18	90/50	Bars and Shapes
				Type XM-19	100/60	Bars and Shapes
				Type XM-29	100/55	Bars and Shapes (Cr-Ni-Mn-N)
		PS	A688	Type XM-29	100/55	Tube (Cr-Ni-Mn-N)
9A	1	PS	A203	Grade A	65/37	Plates (2-1/2 Ni)
				Grade B	70/40	Plates (2-1/2 Ni)
		PS	A234	Grade WPR	63/46	Pipe Fittings
		PS	A333	Grade 7	65/35	Pipe (2-1/2 Ni)
				Grade 9	63/46	Pipe (2 Ni-1 Cu)
		PS	A334	Grade 7	65/35	Tube (2-1/2 Ni)
				Grade 9	63/46	Tube (2 Ni-1 Cu)
		PS	A350	Grade LF9	63/46	Forgings (2 Ni-1 Cu)
		PS	A420	Grade WPL9	63/46	Pipe Fittings (2 Ni-1 Cu)
		PS	A352	Grade LC2	70/40	Casting (2-1/2 Ni)
9B	1	PS	A203	Grade D	65/37	Plates (3-1/2 Ni)
				Grade E	70/40	Plates (3-1/2 Ni)
				Grade F	75/50	Plates (3-1/2 Ni)
		PS	A333	Grade 3	65/35	Pipe (3-1/2 Ni)
		PS	A334	Grade 3	65/35	Tubes (3-1/2 Ni)
		PS	A350	Grade LF3	70/37.5	Forgings (3-1/2 Ni)
		PS	A352	Grade LC3	70/40	Castings (3-1/2 Ni)
		PS	A420	Grade WPL3	65/37	Pipe Fittings (3-1/2 Ni)
		PS	A765	Grade III	70/37.5	Forgings
10A	1	P	A225	Grade C	105/70	Plates (Mn-V)
				Grade D	75/55	Plates (Mn-V)
		PS	A487	Class 1N	85/55	Castings (Mn-V)
				Class 1Q	90/65	Castings (Mn-V)

Table C2 (Continued)

Material No.	Group No.	Std.	Base Metal Specification		Minimum Tensile/Yield, ksi	Type of Base Metal
			Steel and Steel Alloys			
10B	2	PS	A213	Grade T17	60/30	Tubes (1 Cr-V)
10C	3	PS	A612		81/50	Plate (C-Mn-Si)
10E	5	PS	A268	Grade TP446	70/40	Tubes (27 Cr)
				Grade TP329	90/70	Tubes (26 Cr-4 Ni-Mo)
10F	6	PS	A487	Class 2N	85/53	Castings (Mn-Mo)
				Class 2Q	90/65	Castings (Mn-Mo)
				Class 4N	90/60	Castings (Ni-Cr-Mo-V)
10G	7	PS	A658		65/35	Plate (36 Ni)
10H	8	PS	A669		92/64	Tube (18 Cr-5 Ni-3 Mo)
10I	9	PS	A182	Grade FXM-27	60/35	Forging (27 Cr-1 Mo)
		PS	A240	Grade XM-27	65/40	Plate (26 Cr-1 Mo)
				Type XM-33	68/45	Plate (26 Cr-1 Mo-Ti)
				Type XM-33	90/70	Plate (26 Cr-4 Ni-1.5 Mo)
				Type 329	90/70	Plate (26 Cr-4 Ni-1.5 Mo)
		PS	A336	Class FXM-27Eb	60/35	Forgings (26 Cr-1 Mo)
		PS	A479	Grade XM-27	65/40	Bars and Shapes (26 Cr-1 Mo)
		PS	A268	Grade TPXM-27	65/40	Smls. Tubes (26 Cr-1 Mo)
				Grade TPXM-27	65/40	Welded Tubes (26b Cr-1 Mo)
				Grade TPXM-33	68/45	Tubes (26 Cr-1 Mo-Ti)
10J	10	PS	A479	UNS 44700	80/60	Plate (28 Cr-4 Mo)
10K	11	PS	A479	UNS 44800	80/60	Plate (29 Cr-2-1/4 Ni-4 Mo)
11A	1	PS	A333	Grade 8	100/75	Pipe (9 Ni)
11A	1	PS	A334	Grade 8	100/75	Tube (9 Ni)
		PS	A353		100/75	Plate (9 Ni)
		PS	A420	Grade WPL8	100/75	Pipe (9 Ni)
		PS	A522	Type I	100/75	Forgings (*9 Ni)
				Type II	100/75	Forgings (*8 Ni)
		PS	A553	Type I	100/85	Plates (9 Ni)
				Type II	100/85	Plates (8 Ni)
11A	2	PS	A645		96/65	Plates (5 Ni-Mo)
11A	3	PS	A487	Class 4Q	105/85	Castings (Ni-Cr-Mo-V)
				Class 4QA	115/95	Castings (Ni-Cr-Mo)

Table C2 (Continued)

Material No.	Group No.	Std.	Base Metal Specification		Minimum Tensile/Yield, ksi	Type of Base Metal
			Steel and Steel Alloys			
11A	4	PS	A533	Class 3, Grade A	100/82.5	Plate (Mn-Mo)
				Class 3, Grade B	100/82.5	Plate (Mn-Mo-Ni)
				Class 3, Grade C	100/82.5	Plate (Mn-Mo-Ni)
				Class 3, Grade D	100/82.5	Plate (Mn-Mo-Ni)
		PS	A672	Grade J100	100/83	Pipe
11A	5	PS	A508	Class 4	105/85	Forgings (3-1/2 Ni-1-3/4 Cr-1/2 Mo-V)
				Class 4a	115/100	Forgings (3-1/2 Ni-1-3/4 Cr-1/2 Mo-V)
				Class 5	105/85	Forgings (3-1/2 Ni-1-3/4 Cr-1/2 Mo-V)
				Class 5a	115/100	Forgings (3-1/2 Ni-1-3/4 Cr-1/2 Mo-V)
11A	6	PS	A542	Class 1	105/85	Plates (2-1/4 Cr-Mo)
				Class 2	115/100	Plates (2-1/4 Cr-1 Mo)
		S	A543	Type B, Class 1	105/85	Plates (3 Ni-1-3/4 Cr-1/2 Mo)
				Type B, Class 2	115/100	Plates (3 Ni-1-3/4 Cr-1/2 Mo)
				Type B, Class 3	90/70	Plates (3 Ni-1-3/4 Cr-1/2 Mo)
				Type C, Class 1	105/85	Plates (3 Ni-1-1/2 Cr-1/2 Mo)
				Type C, Class 2	115/100	Plates (3 Ni-1-1/2 Cr-1/2 Mo)
				Type C, Class 3	90/70	Plates (3 Ni-1-1/2 Cr-1/2 Mo)
11B	1	S	A514	Grade A	110/100	Plates
	4	S		Grade B	110/100	Plates
				Grade C	110/100	Plates
	5	S		Grade D	110/100	Plates
	2	S		Grade E	110/100	Plates
	3	S		Grade F	110/100	Plates
				Grade G	110/100	Plates
				Grade H	110/100	Plates
	6	S		Grade J	110/100	Plates

Table C2 (Continued)

Material No.	Group No.	Std.	Base Metal Specification		Minimum Tensile/Yield, ksi	Type of Base Metal
				Steel and Steel Alloys		
11B			A514	Grade K	110/100	Plates
				Grade L	110/100	Plates
				Grade M	110/100	Plates
				Grade N	110/100	Plates
	8	S		Grade P	110/100	Plates
	9	S		Grade Q	110/100	Plates
		PS	A517	Grade A	115/100	Plates
		S		Grade G	115/100	Alloy Steel Plates
		S		Grade H	115/100	Alloy Steel Plates
		S		Grade J	115/100	Alloy Steel Plates
		S		Grade K	115/100	Alloy Steel Plates
		S		Grade L	115/100	Alloy Steel Plates
		S		Grade Q	115/100	Alloy Steel Plate
		S	A519	Grade 4130	75/55	Mechanical Tubing
				Grade 9630	75/55	Mechanical Tubing
		PS	A592	Grade A	115/100	Alloy Steel Forgings
		S	A709	Grade 100	100/100	Structural Steel
				Grade 100W	110/100	Structural Steel
		S	A513	Grade 4130	*	Mechanical Tubing
				Grade 8630	*	Mechanical Tubing
11B	2	PS	A513	Grade E	115/100	Alloy Steel Forgings
		PS	A517	Grade E	116/100	Plates
		PS	A592	Grade E	115/100	Alloy Steel Forgings
11B	3	PS	A517	Grade F	115/100	Plates
		PS	A592	Grade F	115/100	Alloy Steel Forgings
11B	4	PS	A517	Grade B	115/100	Plates
		S		Grade C	115/100	Plates
	5	PS		Grade D	115/100	Plates
	6	PS		Grade J	115/100	Plates
				Grade M	115/100	Plates
	8	PS		Grade P	115/100	Plates

Table C2 (Continued)

Material No.	Std.	Base Metal Specification	Thickness, in.	Minimum Tensile/Yield, ksi	Type of Base Metal	Notes
			Aluminum and Aluminum-Base Alloys			
21	PS	B209	0.051-3.000	8/2.5	1060 Sheet, Plate	
	PS	B209	0.051-3.000	11/3.5	1100 Sheet, Plate	
	PS	B209	0.051-3.000	14/5.5	3003 Sheet, Plate	
	PS	B209	0.051-0.499	13/4.5	Alclad 3003 Sheet, Plate	(2)
			0.500-3.000	14/5.0	Alclad 3003 Plate	(3)
	PS	B210	All	8.5/2.5	1060 Bars, Rods, Shapes, Tubes	(1)
	PS	B221	All	8.5/2.5	1060 Bars, Rods, Shapes, Tubes	(1)
	PS	B221	All	11/3	1100 Bars, Rods, Shapes, Tubes	
	PS	B221	All	14/5	3003 Bars, Rods, Shapes, Tubes	(1)
	PS	B234	0.010-0.200	12/10	1060 Seamless Pipe, Tube	
	PS	B241	All	11/3	1100 Bars, Rods, Shapes, Tubes	(1)
	PS	B241	All	8.5/2.5	1060 Seamless Pipe, Tube	
			All	11/3	1100 Seamless Pipe, Tube	
			All	19/5	3003 Seamless Pipe, Tube	
	PS	B247	Up thru 4.000	14/5	3003 Die Forgings	
	PS	B210	All	14/5	3003 Tube	
22	PS	B209	0.050-3.000	22/8.5	3004 Sheet, Plate	
	PS	B209	0.051-0.499	21/8.0	3004 Sheet, Plate	(2)
	PS	B209	0.500-3.000	22/8.5	Alclad 3004 Plate	(3)
	PS	B209	0.051-3.000	25/9.5	5052 Sheet, Plate	
	PS	B209	0.051-3.000	30/11	5254 Sheet, Plate	
	PS	B209	0.020-3.000	30/11	5154 Sheet, Plate	
	PS	B209	0.020-3.000	31/12	5454 Sheet, Plate	
	PS	B209	0.051-3.000	25/9.5	5652 Sheet, Plate	
	PS	B210	0.018-0.450	25/10	5052 Seamless Tube	
	PS	B210	All	30/11	5154 Bars, Rods, Shapes, Tube	(1)
	PS	B221	All	30/11	5154 Bars, Rods, Shapes, Tube	
	PS	B221	All	31/12	5454 Bars, Rods, Shapes, Tube	
	PS	B234	All	31/12	5454 Bars, Rods, Shapes, Tube	(1)
	PS	B234	All	30/12	5154 Bars, Rods, Shapes, Tube	
	PS	B241	All	31/12	5052 Tube	

Notes:
1. Some of the indicated product forms are not normally produced in all sizes indicated; for more specific coverage, see applicable tables of ASME Boiler and Pressure Vessel Code, Section II, Part B.
2. Specified tensile properties are for full-thickness specimens which include cladding.
3. Specified tensile properties are for specimens taken from the core.

Table C2 (Continued)

Material No.	Std.	Base Metal Specification	Thickness, in.	Minimum Tensile/Yield, ksi	Type of Base Metal	Notes
			Aluminum and Aluminum-Base Alloys			
23	PS	B209	0.051-6.000	24/12	6061 Sheet, Plate (T4, T6 welded)	
	PS	B209	0.051-5.000	24/14	Alclad 6061 Sheet, Plate (T4, T6 welded)	
	PS	B210	All	24/14	6061 Bars, Rods, Shapes, Pipe, Tube (T4, T6 welded)	(1)
	PS	B210	All	17/8	6063 Bars, Rods, Shapes, Tubes (T4, T6 welded)	(1)
	PS	B211	All	24/14	6061 Bars, Rods, Shapes, Pipe, Tube (T4, T6 welded)	(1)
	PS	B221	All	17/8	6063 Bars, Rods, Shapes, Pipe, Tube (T4, T6 welded)	(1)
	PS	B241	All	24/16	6061 Pipe, Tube (T4, T6 welded)	
	PS	B247	Up Thru 8.000	37/34	6061 Forgings (T6 welded)	
	PS	B308	All	38/35	6061 Structural Shapes (T6 welded)	
25	PS	B209	0.051-1.500	40/18	5083 Sheet, Plate	
			1.501-3.000	39/17	5083 Plate	
			3.001-5.000	38/16	5083 Plate	
			5.001-7.000	37/15	5083 Plate	
			7.001-8.000	36/14	5083 Plate	
	PS	B209	0.020-2.000	34/14	5086 Plate, Sheet	
	PS	B209	0.051-1.500	42/19	5456 Sheet, Plate	
			1.501-3.000	41/18	5456 Plate	
			3.001-5.000	40/17	5456 Plate	
			5.001-7.000	39/16	5456 Plate	
			7.001-8.000	38/15	5456 Plate	
	PS	B221	Up Thru 5.000	39/16	5083 Bars, Rods, Shapes, Tube	
	PS	B241	Up Thru 5.000	35/14	5086 Smls. Pipe and Tube	
	PS	B241	Up Thru 5.000	41/19	5456 Bars, Rods, Shapes, Tube	
	PS	B241	Up Thru 5.000	39/16	5083 Bars, Rods, Shapes, Tube	
	PS	B247	Up Thru 4.000	39/16	5083 Forgings	
	PS	B221	Up Thru 5.000	41/19	5456 Bars, Rods, Shapes, Tube	

Note:

1. Some of the indicated product forms are not normally produced in all sizes indicated; for more specific coverage, see applicable tables of ASME Boiler and Pressure Vessel Code, Section II, Part B.

Table C2 (Continued)

Material No.	Std.	Base Metal Specification	Condition	Size(s) or Thickness	Minimum Tensile/Yield, ksi	Type of Base Metal
			Copper and Copper-Base Alloys			
31	PS	B42	Drawn	1/8-2	45/40	C10200, C12000, C12200 Pipe
			Drawn	2-1/2-12	36/30	
	PS	B75	Annealed	---	30/9	C10200, C12000, C12200, C14200 Smls. Tube
	PS	B111	Light Drawn	---	36/30	C10200, C12000, C12200, C14200
			Hard Drawn	---	45/40	Smls. Tube
		B111	Annealed	---	38/12	C19200 Smls. Tube
			Hard Drawn	---	48/43	C19200 Smls. Tube
	S	B133	Annealed	---	30/-	C10200, C11000, C12000, C12200, C14200 Rod
	PS	B152	Annealed (Phos. Dox.)	---	30/10	C10200, C10400, C10500, C10700, C12200, C12300, Sheet, Plate Bar
	PS	B395	Light Drawn	---	36/30	C10200, C12000, C12200, C14200
		B395	Annealed	---	38/12	C19200 Tube
	PS	B543	Annealed	---	32/15	C12200 Welded Tube
	PS	B543	Annealed	---	45/22	C19400 Welded Tube
			Welded from Annealed Strip			C19400 Welded Tube
32	PS	B21	Extruded	---	50/20	C46200
	PS	B21	Soft	---	50/20	C46400
	PS	B43	Annealed	---	40/12	Brass Tube, Pipe
	PS	B111	Annealed	---	45/15	C44300 Admiralty, Tube
	PS	B111	Annealed	---	45/15	C44400 Admiralty, Tube
	PS	B111	Annealed	---	45/15	C44500 Admiralty, Tube
	PS	B111	Annealed	---	50/20	C28000 Muntz Metal, Tube
	PS	B111	Annealed	---	40/12	C23000 red
	PS	B111	Annealed	---	50/18	C68700 Seamless Tube
	PS	B135	Annealed	---	40/12	Seamless Brass Tube
	PS	B171	Annealed	4 and under	45/15	C44300, C44400, C44500 Plate
	PS	B171	Annealed	To 5	50/18	C46400 Naval Brass, Plate

Table C2 (Continued)

Material No.	Std.	Base Metal Specification	Condition	Size(s) or Thickness	Minimum Tensile/Yield, ksi	Type of Base Metal
			Copper and Copper-Base Alloys			
32	PS	B171	Annealed	2 and under	50/20	C36500
			Annealed	Over 2 thru 3.5	45/15	Muntz Metal Plate
			Annealed	Over 3.5 thru 5	40/12	
	PS	B359	Annealed	---	40/12	C2300 On Seamless Tube
	PS	B359	Annealed	---	45/15	C44300, C44400, C44500
	PS	B359	Annealed	---	50/18	C68700 Seamless Tube
	PS	B543	Annealed	---	40/12	C23000 Welded Tube
		B543	Annealed	---	45/15	C44300, C44400, C44500 Welded Tube
		B543	Annealed	---	50/18	C68700 Welded Tube
33	PS	B96	Annealed	---	50/18	C65500 Cu-Si Plate, Sheet
	PS	B98	Soft	---	52/15	C65500 Cu-Si Rod, Bar, Shape
	PS	B98	Soft	---	40/12	C65100 Cu-Si Rod, Bar, Shapes
	PS	B315	Annealed	---	50/15	C65500 Cu-Si Pipe and Tube
34	PS	B111	Annealed	---	52/18	C71500 Cu-Ni Tube
	PS	B111	Light Drawn	---	40/30	C70400 Smls. Tube
	PS	B111	Annealed	---	40/15	C70600 Cu-Ni Tube
	PS	B171	Annealed	2.5 and under	50/20	C71500 Cu-Ni Plate
	PS	B171	Annealed	2.5 and under incl.	40/15	C70600 Cu-Ni Plate
		B402	Annealed	Over 2.5 to 5 incl.	45/18	C71500 Cu-Ni Plate
		B402	Annealed	---	40/15	C70600
		B402	Annealed	---	50/20	C71500
		B402	Annealed	---	42/16	C72200
	PS	B359	Annealed	---	38/12	C70400 Smls. Tube
		B359	Annealed	---	40/15	C70600 Smls. Tube
		B359	Annealed	---	45/16	C71000 Smls. Tube
		B359	Annealed	---	52/18	C71500 Smls. Tube
	PS	B395	Annealed	---	40/15	C70600 Smls. Tube

Table C2 (Continued)

Material No.	Std.	Base Metal Specification	Condition	Size(s) or Thickness	Minimum Tensile/Yield, ksi	Type of Base Metal
			Copper and Copper-Base Alloys			
34	PS	B395	Annealed	---	45/16	C71000 Smls. Tube
	PS	B395	Annealed	---	52/18	C71500 Smls. Tube
	PS	B395	Drawn, Stress Relieve	---	72/50	C71500 Smls. Tube
	PS	B466	Annealed	---	50/18	C71500 (30 Ni) Cu-Ni Pipe, Tube
	PS	B466	Annealed	---	45/16	C71000 (20 Ni) Cu-Ni Pipe, Tube
	PS	B466	Annealed	---	38/13	C70600 (10 Ni) Cu-Ni
	PS	B467	Annealed	To 4-1/2	50/20	C71500 (30 Ni) Cu-Ni Pipe, Tube
				Over 4-1/2	45/15	C71500 (30 Ni) Cu-Ni Pipe, Tube
	PS	B467	Annealed	To 4-1/2	40/15	C70600 (10 Ni) Cu-Ni Pipe, Tube
				Over 4-1/2	38/13	C70600 (10 Ni) Cu-Ni Pipe, Tube
	PS	B543	Annealed	---	38/12	C70400 Welded Tube Cu-Ni
	PS	B543	Annealed	---	40/15	C70600 Welded Tube Cu-Ni
	PS	B543	Welded From Annealed Strip	---	45/30	C70600 Welded Tube Cu-Ni
	PS	B543	Annealed	---	52/18	C71500 Welded Tube Cu-Ni
		B543	Hard Drawn	---	72/50	C71500 Welded Tube Cu-Ni
35	PS	B148	As Cast	---	65/25	C95200 Aluminum Bronze Casting
	PS	B148	As Cast	---	75/30	C95400 Aluminum Bronze Casting
	PS	B148	Heat Treated	---	90/45	C95400 Al-Bronze Casting
	PS	B150	Annealed	1/2 in. and under	80/40	C61400 Rod, Bar and Shapes
				1/2 to 1 in.	75/35	
				Over 1 in. to 3 in.	70/32	
		B150	Annealed	1/2 to 1 in. Over 1 in. to 2 in.	100/50	C63000 Rod, Bar, and Shapes
				2 in. to 4 in.	85/42.5	
				Over 4 in.	80/40	

Table C2 (Continued)

Material No.	Std.	Base Metal Specification	Condition	Size(s) or Thickness	Minimum Tensile/Yield, ksi	Type of Base Metal
			Copper and Copper-Base Alloys			
35	PS	B150	Annealed	1/2 in. and under	90/45	C64200 Rod, Bar, and Shapes
				1/2 in. to 1 in.	85/45	
				Over 1 in. to 2 in.	80/42	
				Over 2 in. to 3 in.	75/35	
	PS	B169	Annealed	---	50/20	C61000 Aluminum Bronze Plate Sheet
	PS	B169	Annealed	1/2 and under	72/32	C61400 Aluminum Bronze Plate Sheet
			Annealed	Over 1/2 to 2	70/30	
			Annealed	2 to 5 incl.	65/28	
	PS	B171	Annealed	2 and under	70/30	C61400
				2 to 5 incl.	65/28	Aluminum Bronze, Plate
	PS	B171	Annealed	2 and under	90/36	C63000
				Over 2 to 3.5 incl.	85/33	Aluminum Bronze, Plate
				Over 3.5 to 5 incl.	80/30	
	PS	B271	As Cast	---	65/25	C95200 Aluminum Bronze Casting
35	PS	B111	Annealed	---	50/19	C60800 Al Bronze Tube

Table C2 (Continued)

Material No.	Std.	Base Metal Specification	Alloy Designation	Minimum Tensile/ Yield, ksi	Type of Base Metal
		Nickel and Nickel-Base Alloys			
41	PS	B-160 Annealed	N02200	55/15	Rod, Bar
	PS	B-160 Hot Rolled	N02200	60/15	Rod, Bar
	PS	B-160 Annealed	N02201	50/10	Low C, Ni Rod, Bar
	PS	B-160 Hot Rolled	N02201	50/10	Low C, Ni Rod, Bar
	PS	B-161 Annealed	N02200	55/15	Pipe, Tube (99.0 Ni)
	PS	B-161 Stress Relieved	N02200	65/40	Pipe, Tube (99.0 Ni)
	PS	B-161 Stress Relieved	N02201	60/30	Low C, Ni Pipe/Tube
	PS	B-161 Annealed	N02201	50/12	Low C, Ni Pipe/Tube
	PS	B-162 Annealed	N02200	55/15	Plate, Sheet, Strip (99.0 Ni)
	PS	B-162 Hot Rolled	N02200	55/15	Plate, Sheet, Strip (99.0 Ni)
	PS	B-162 Annealed	N02201	50/12	Low C, Ni Plate, Sheet, Strip
	PS	B-162 Hot Rolled	N02201	50/12	Low C, Ni Plate, Sheet, Strip
	PS	B-163 Annealed	N02200	55/15	Condenser Tube (99.0 Ni)
	PS	B-163 Stress Relieved	N02200	65/40	Condenser Tube (99.0 Ni)
	PS	B-163 Annealed	N02201	50/12	Low C, Ni Condenser Tube (99.0 Ni)
	PS	B-163 Stress Relieved	N02201	60/30	Low C, Ni Condenser Tube (99.0 Ni)
42	PS	B-127 Annealed	N04400	70/28	Ni-Cu Plate, Sheet, Strip
	PS	B-127 Hot Rolled	N04400	75/40	Ni-Cu Plate, Sheet, Strip
	PS	B-163 Annealed	N04400	70/28	Ni-Cu Condenser Tube
	PS	B-163 Stress Relieved	N04400	85/55	Ni-Cu Condenser Tube
	PS	B-164 Annealed	N04400	70/25	Ni-Cu Rod, Bar
	PS	B-164 Hot Finished	N04400	75/30	Ni-Cu Rod, Bar
	PS	B-164 Stress Relieved 4″ to 12″ Dia.	N04400	75/30	Ni-Cu Rod, Bar
		Incl.	N04400	80/40	Ni-Cu Rod, Bar
	PS	B-164 Stress Relieved over 12″ Dia.	N04400	75/40	Ni-Cu Rod, Bar
	PS	B-165 Annealed	N04400	70/28	Ni-Cu Pipe, Tube
	PS	B-165 Stress Relieved	N04400	85/55	Ni-Cu Pipe, Tube
	PS	B-564 Annealed	N04400	70/28	Ni-Cu Forgings
43	PS	B-163 Annealed	N06600	80/35	Ni-Cr-Fe Condenser Tube
	PS	B-166 Annealed	N06600	80/35	Ni-Cr-Fe Rod, Bar

Table C2 (Continued)

Material No.	Std.	Base Metal Specification	Alloy Designation	Minimum Tensile/ Yield, ksi	Type of Base Metal
		Nickel and Nickel-Base Alloys			
43	PS	B-166 Hot Finished 1/4″-3″ incl.	N06600	90/40	Ni-Cr-Fe Rod, Bar
	PS	B-166 Hot Finished over 3″ and Hex.	N06600	85/35	Ni-Cr-Fe Rod, Bar
	PS	B-167 C D/Annealed 5″ and under	N06600	80/35	Ni-Cr-Fe Pipe, Tube Annealed
	PS	B-167 C D/Annealed over 5″ O.D.	N06600	80/30	Ni-Cr-Fe Pipe, Tube
	PS	B-167 Hot Finished/ Annealed 5″ O.D. and under		80/30	Ni-Cr-Fe Pipe, Tube
	PS	B-167 Hot Finished/ Annealed over 5″ O.D.	N06600	75/25	Ni-Cr-Fe Pipe, Tube
	PS	B-168 Annealed	N06600	80/35	Plate, Sheet, Strip
	PS	B-168 Hot Rolled	N06600	85/35	Plate, Sheet, Strip
	PS	B-443 Annealed	N06625	120/60	Plate, Sheet, Strip Ni-Cr-Mo-Cb
	PS	B-444 Annealed	N06625	120/60	Pipe, Tube Ni, Cr-Mo-Cb
	PS	B-446 Annealed	N06625	120/60	Rod, Bar Ni-Cr-Mo-Cb
	PS	B-435 under 3/16″	N06002	100/40	Sheet Ni-Cr-Mo-Fe
	PS	B-435 3/16″ and over	N06002	95/35	Plate, Ni-Cr-Mo-Fe
	PS	B517	N06600	80/35	Pipe Ni-Cr-Fe
	PS	B-564	N06002	80/35	Ni-Cu-Fe Forgings
	PS	B-572	N06002	95/35	Rod Ni-Cr-Mo-Fe
	PS	B-619 Annealed	N06002	100/40	Welded Pipe Ni-Cr-Mo-Fe
	PS	B-622 Annealed	N06002	100/40	Seamless Pipe and Tube Ni-Cr-Mo-Fe
	PS	B-626 Annealed	N06002	100/40	Tube Ni-Cr-Mo-Fe
44	S	B-333 Annealed under 3/16″	N10001	115/50	Sheet Ni-Mo
	PS	B-333 Annealed 3/16-2-1/2″ incl.	N10001	100/45	Sheet Ni-Mo
	S	B-333 Annealed	N10665	110/51	Plate, Sheet, Strip Ni-Mo
	PS	B-335 Annealed 5/16-1-1/2″ incl.	N10001	115/46	Rod, Bar Ni-Mo
	PS	B-335 Annealed over 1-1/2-3-1/2″ incl.	N10001	100/46	Rod, Bar Ni-Mo

Table C2 (Continued)

Material No.	Std.	Base Metal Specification	Alloy Designation	Minimum Tensile/ Yield, ksi	Type of Base Metal
		Nickel and Nickel-Base Alloys			
44	PS	B-335 Annealed	N10665	110/51	Rod Ni-Mo
	PS	B-434 Annealed	N10003	100/40	Plate, Sheet Ni-Mo-Cr
	PS	B-573 Annealed	N10003	100/40	Rod Ni-Mo-Cr-Fe
	PS	B-574 Annealed	N10276	100/41	Rod Low C-Ni-Mo-Cr
	PS	B-574 Annealed	N06455	100/40	Rod Low C-Ni-Mo-Cr
	PS	B-575 Annealed	N10276	100/41	Plate, Sheet, Strip Low
	PS	B-575 Annealed	N06455	100/40	Plate, Sheet, Strip Low C-Ni-Mo-Cr
	PS	B-619 Annealed	N10001	100/45	Welded Pipe Ni-Mo
	PS	B-619 Annealed	N10665	110/51	Welded Pipe Ni-Mo
	PS	B-619 Annealed	N10276	100/41	Welded Pipe Low C-Ni-Mo-Cr
	PS	B-619 Annealed	N06455	100/40	Welded Pipe Low C-Ni-Mo-Cr
	PS	B-622 Annealed	N10001	100/45	Smls. Pipe and Tube Ni-Mo
	PS	B-622 Annealed	N10065	110/51	Smls. Pipe and Tube Ni-Mo
	PS	B-622 Annealed	N10276	100/41	Smls. Pipe and Tube Low C-Ni-Mo-Cr
	PS	B-622 Annealed	N06455	100/40	Smls. Pipe and Tube Low C-Ni-Mo-Cr
	PS	B-626	N010001	100/45	Tube Ni-Mo
	PS	B-626	N010665	110/51	Tube (28 Mo)
	PS	B-626	N06455	100/40	Tube (15 Mo-16 Cr)
	PS	B-626	N10276	100/41	Tube (16 Mo)
45	PS	B-163 Annealed Gr. 1	N08800	75/30	Condenser Tube Ni-Cr-Fe
	PS	B-163 Annealed Gr. 2	N08810	65/25	Condenser Tube Ni-Fe-Cr
	S	B-167 Hot Finished Annealed	N06600	80/30	Smls. Pipe and Tube Ni-Cr-Fe
	PS	B-407 Annealed	N08800	75/30	Pipe, Tube Ni-Cr-Fe
	PS	B-407 Annealed	N08800	65/25	Pipe, Tube Ni-Cr-Fe
	PS	B-408 Annealed	N08800	75/30	Rod, Bar Ni-Fe-Cr
	PS	B-408 Annealed	N08810	65/25	Rod, Bar Ni-Fe-Cr
	PS	B-409 Annealed	N08800	75/30	Plate, Sheet, Strip Ni-Fe-Cr
	PS	B-409 Annealed	N08810	65/25	Plate, Sheet, Strip Ni-Fe-Cr
	PS	B-462 Annealed	N08020	80/35	Forgings (Ni-Fe-Cr-Cb)

Table C2 (Continued)

Material No.	Std.	Base Metal Specification	Alloy Designation	Minimum Tensile/ Yield, ksi	Type of Base Metal
		Nickel and Nickel-Base Alloys			
45	PS	B-463 Annealed	N08825	85/35	Condenser Tube Ni-Fe-Cr-Mo-Cu
	PS	B-423 Annealed	N08825	75/25	Pipe, Tube Ni-Fe-Cr-Mo-Cu
	PS	B-424 Annealed	N08825	85/35	Plate, Sheet, Strip Ni-Fe-Cr-Mo-Cu
	PS	B-425 Annealed	N08825	85/35	Rod, Bar Ni-Cr-Mo-Cu
	PS	B-463 Annealed	N08020	85/40	Plate, Sheet, and Strip Cr-Ni-Fe-Mo-Cu-Cb
	PS	B-464 Seamless Annealed & Welded Annealed	N08020	85/40	Pipe/Tube Cr-Ni-Fe-Mo-Cu-Cb
	PS	B-468 Seamless Annealed & Welded Annealed	N08020	85/40	Tubes Cr-Ni-Fe-Mo-Cu-Cb
	PS	B-473 Annealed	N08020	85/35	Bars/Rods/Shapes/Forgings Cr-Ni-Fe-Mo-Cu-Cb
	PS	B-516	N06600	80/35	Welded Pipe Ni-Fe-Cr-Mo-Cu
	PS	B-517	N06600	80/35	Welded Tube Ni-Cr-Fe-Mo-Cu
	PS	B-564 Annealed	N08800	75/30	Forgings Ni-Fe-Cr
	PS	B-564 Annealed	N08810	65/25	Forgings Ni-Fe-Cr
	PS	B-581 Up to 3/4" incl.	N06007	90/35	Rod Ni-Cr-Fe-Mo-Cu
	PS	B-581 5/16 in. to 3-1/2 in.	N06975	95/30	Rod Ni-Cr-Fe-Mo-Cu
	PS	B-581 Above 3/4 in.	N06007	85/30	Rod Ni-Cr-Fe-Mo-Cu
	PS	B-582 Up to 3/4 in.	N06007	90/35	Plate, Sheet, Strip Ni-Cr-Fe-Mo-Cu
	PS	B-582 Above 3/4 in. to 2-1/2 in.	N06007	85/30	Plate, Sheet, Strip Ni-Cr-Fe-Mo-Cu
	PS	B-619 Annealed	N08320	75/28	Welded Pipe Ni-Fe-Cr-Mo
	PS	B-619 Annealed	N06007	90/35	Welded Pipe Ni-Cr-Fe-Mo-Cu
	PS	B-619 Annealed	N06975	85/32	Pipe Ni-Cr-Fe-Mo-Cu
	PS	B-620 Annealed	N08320	75/28	Sheet, Plate, Strip Ni-Fe-Cr-Mo
	PS	B-621 Annealed	N08320	75/28	Bar Ni-Fe-Cr-Mo
	PS	B-622 Annealed	N08320	75/28	Smls. Pipe & Tube Ni-Fe-Cr-Mo
	PS	B-622 Annealed	N06007	90/35	Smls. Pipe & Tube Ni-Cr-Fe-Mo-Cu
	PS	B-625	N08904	71/31	Plate, Sheet, Strip Ni-Fe-Cr-Mo-Cu
	PS	B-626	N06007	90/35	Welded Tube Ni-Cr-Fe-Mo-Cu
	PS	B-626	N08320	75/28	Welded Tube Ni-Fe-Cr-Mo
	PS	B-626	N06975	85/32	Welded Tube Ni-Cr-Fe-Mo-Cu

Table C2 (Continued)

Material No.	Std.	Base Metal Specification		Minimum Tensile/ Yield, ksi	Type of Base Metal
		Nickel and Nickel-Base Alloys			
46	PS	B-511	Annealed N08330	70/30	Bar Ni-Fe-Cr-Si
	PS	B-535	Annealed N08330	70/30	Smls. & Welded Pipe Ni-Fe-Cr-Si
	PS	B-536	Annealed N08330	70/30	Plate, Sheet, Strip Ni-Fe-Cr-Si
		Titanium and Titanium-Base Alloys			
51	PS	B-265	Grade 1	35/25	Strip, Sheet & Plate
			Grade 2	50/40	Strip, Sheet & Plate
			Grade 7	50/40	Alloy, Strip, Sheet, Plate
	PS	B-337	Grade 1	35/25	Smls. & Welded Pipe
			Grade 2	50/40	Smls. & Welded Pipe
			Grade 7	50/40	Smls. & Welded Pipe
	PS	B-338	Grade 1	35/25	Smls. & Welded Pipe
			Grade 2	50/40	Smls. & Welded Pipe
			Grade 7	50/40	Smls. & Welded Pipe
	PS	B-348	Grade 1	35/25	Bars & Billets
			Grade 2	50/40	Bars & Billets
	PS		Grade 7	50/40	Alloy Bars and Billets
	PS	B-381	Grade F-1	35/25	Forgings
			Grade F-2	50/40	Forgings
52	PS	B-265	Grade 3	65/55	Strip, Sheet & Plate
			Grade 12	70/50	Strip, Sheet & Plate
	PS	B-337	Grade 3	65/55	Seamless & Welded Pipe
			Grade 12	70/50	Seamless & Welded Pipe
	PS	B-338	Grade 3	65/55	Seamless & Welded Pipe
			Grade 12	70/50	Seamless & Welded Pipe
	PS	B-348	Grade 3	65/55	Bars & Billets
			Grade 12	70/50	Bars & Billets
	PS	B-381	Grade F-3	65/55	Forgings
			Grade F-12	70/50	Forgings
		Zirconium and Zirconium-Base Alloys			
61	PS	B-493	Grade R60702	52/32	Forgings and Extrusions
	PS	B-523	Grade R60702	55/30	Seamless and Welded Tubing
	PS	B-550	Grade R60702	55/30	Rods & Bars
	PS	B-551	Grade R60702	52/30	Strip, Sheet & Plate

Appendix D
Radiographic Examination of Welder Performance Qualification Test Weldments

D1. General

This procedure defines requirements for radiographic examination of welder performance qualification test weldments where called for in Section 3.

D2. Radiographic Procedure

D2.1 Radiographs shall be made by either x-rays or other high energy radiation source methods. Film shall be clean and free of artifacts and film processing defects in the weld area, heat-affected zone, penetrameter, and identification locations. Radiographic procedures should be performed according to ASME Section V, Article 2. Final acceptance of the film for viewing shall be based upon the ability to see the prescribed penetrameter image and the specified hole. Composite viewing of film is not permitted. Radiographs shall show:

D2.1.1 The required hole in each penetrameter specified by Table D1 or D2, as applicable.

D2.1.2 The penetrameter identification number.

D2.1.3 The welder's name or identification number, weld procedure, and test weldment position.

D2.1.4 The radiograph(s) shall show the entire weld area and be within the required film density. When multiple radiographs are required to cover the weld, location markers shall be used to demonstrate complete weld coverage.

D2.2 Radiography shall be performed in accordance with all applicable safety requirements.

D2.3 Penetrameters shall be selected from Table D1 for single-wall radiography and from Table D2 for double-wall radiography, using either single-wall or double-wall viewing. When film side penetrameters are used, a lead letter "F" shall be placed adjacent to the penetrameter. Penetrameter selection shall be based on the actual wall thickness plus reinforcement. A minimum of one penetrameter is required for each radiograph. Penetrameters shall be placed on the side of the work nearest the radiation source, when possible. Penetrameters shall conform to the requirements of ASTM E142 in design, group, and grade. Alternative wire penetrameters that may be used are provided for in ASTM E747, Controlling Quality of Radiographic Testing Using Wire Penetrameters.

D2.4 Face reinforcement may be removed at the option of the qualifier. Root reinforcement shall not be removed from single-walled groove joints. Backing, if used, shall not be removed. Shims of radiographically similar material shall be placed under the penetrameter so that the total thickness of material between the penetrameter and the film is at least equal to the average thickness of the weld measured through its reinforcement and backing.

D2.5 Radiographs shall be made with a single source of radiation approximately centered over the weld and the length of weld or segment being examined. The perpendicular distance from the radiation source to the film shall not be less than seven times the maximum thickness of the weld under examination. The film, during exposure, shall be as close to the surface of the weld opposite the source of radiation as possible.

D3. Personnel Requirements

Personnel performing radiography shall be a minimum of ASNT Level I or equivalent. Interpreting of radiographs shall be the responsibility of the qualifier.

Table D1
Single-wall radiographic technique

Metal thickness plus reinforcement, in.	Penetrameter			
	Source side		Film side	
	Designation	Essential hole	Designation	Essential hole
Up to 0.25 incl.	10	4T	7	4T
Over 0.25 thru 0.375	12	4T	10	4T
Over 0.375 thru 0.50	15	4T	12	4T
Over 0.50 thru 0.625	15	4T	12	4T
Over 0.625 thru 0.75	17	4T	15	4T
Over 0.75 thru 0.875	20	4T	17	4T
Over 0.875 thru 1.00	20	4T	17	4T
Over 1.00 thru 1.25	25	4T	20	4T
Over 1.25 thru 1.50	30	2T	25	2T
Over 1.50 thru 2.00	35	2T	20	2T
Over 2.00 thru 2.50	40	2T	35	2T
Over 2.50 thru 3.00	45	2T	40	2T
Over 3.00 thru 4.00	50	2T	45	2T
Over 4.00 thru 6.00	60	2T	50	2T
Over 6.00 thru 8.00	80	2T	60	2T
Over 8.00 thru 10.00	100	2T	80	2T
Over 10.00 thru 12.00	120	2T	100	2T
Over 12.00 thru 16.00	160	2T	120	2T
Over 16.00 thru 20.00	200	2T	160	2T

Table D2
Double-wall radiographic technique

Metal thickness plus reinforcement, in.	Film or source side penetrameter	
	Designation	Essential hole
0 thru 0.375	10	4T
Over 0.357 thru 0.625	12	4T
Over 0.625 thru 0.875	15	4T
Over 0.875 thru 1.00	17	4T
Over 1.00 thru 1.50	25	2T
Over 1.50 thru 2.50	30	2T
Over 2.50 thru 3.00	35	2T
Over 3.00 thru 4.00	40	2T
Over 4.00 thru 6.00	50	2T

SI Conversions for Tables D1 and D2

in.	mm	in.	mm
0.25	6.3	2.00	50.8
0.357	9.52	2.50	63.5
0.50	12.7	3.00	76.2
0.625	15.87	4.00	101.6
0.75	19.0	6.00	152.4
0.875	22.22	8.00	203.2
1.00	25.4	10.00	254.0
1.25	31.7	12.00	304.8
1.50	38.1	16.00	406.4
		20.00	508.0

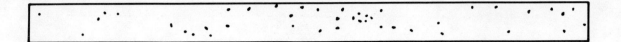

Typical quantity and size permitted in 6-in. length of weld 1/8 in. to 1/4 in. thickness

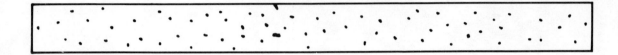

Typical quantity and size permitted in 6-in. length of weld over 1/4 in. to 1/2 in. thickness

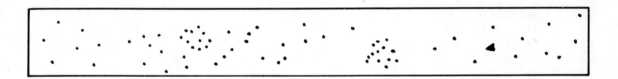

Typical quantity and size permitted in 6-in. length of weld over 1/2 in. to 1 in. thickness

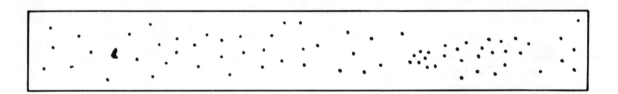

Typical quantity and size permitted in 6-in. length of weld over 1 in. thickness

Fig. D1 — Rounded indication charts

Appendix E
Macroetch Procedure

E1. General

This procedure sets forth requirements for examination of weld specimens by macroetching where called for in Sections 2 or 3. The surfaces to be etched should be smoothed by filing, machining, or grinding on metallographic papers. With different alloys and tempers, the etching period will vary from a few seconds to several minutes, and should be continued until the desired contrast is obtained. As a protection from the fumes liberated during the etching process, this work should be done under a hood. After etching, the specimens should be thoroughly rinsed and then dried with a blast of warm air. Coating the surface with a thin clear lacquer will preserve the appearance.

E2. Etching Solutions and Procedures

Caution: (See E3 for safety precautions).

E2.1 Ferrous Metal. Etching solutions suitable for carbon and low-alloy steels, together with directions for their use, are suggested as follows (other solutions may be used):

E2.1.1 Hydrochloric Acid. Hydrochloric (muriatic) acid and water, equal parts by volume. The solution should be kept at or near the boiling temperature during the etching process. The specimens are to be immersed in the solution for a sufficient period of time to reveal all lack of soundness that might exist at their cross-sectional surfaces.

E2.1.2 Ammonium Persulfate. One part of ammonium persulfate to nine parts of water by weight. The solution should be used at room temperature, and should be applied by vigorously rubbing the surface to be etched with a piece of cotton saturated with the solution. The etching process should be continued until there is a clear definition of the macrostructure in the weld.

E2.1.3 Iodine and Potassium Iodide. One part of powdered iodine (solid form), two parts of powdered potassium iodide, and ten parts of water, all by weight. The solution should be used at room temperature, and brushed on the surface to be etched until there is a clear definition or outline of the weld.

E2.1.4 Nitric Acid. One part of nitric acid and three parts of water, by volume. (**Caution:** Always pour the acid into the water. Nitric acid causes bad stains and severe burns.) The solution may be used at room temperature and applied to the surface to be etched with a glass stirring rod. The specimens may also be placed in a boiling solution of the acid, but the work should be done in a well ventilated room. The etching process should be continued for a sufficient period of time to reveal all lack of soundness that might exist at the cross-sectional surfaces of the weld.

E2.2 Nonferrous Metals. The following etching reagents and directions for their use are suggested for revealing the macrostructure:

E2.2.1 Aluminum and Aluminum-Base Alloys. The solution given below is to be used at room temperature, and etching is accomplished by either swabbing or immersing the specimen.

Hydrochloric acid (conc.)	15 ml
Hydrofluoric acid (48%)	10 ml
Water 85	ml

E2.2.2 Copper and Copper-Base Alloys. Cold Concentrated Nitric Acid. Etching is accomplished by either flooding or immersing the specimen for several seconds under a hood. After rinsing with a flood of water, the process is repeated with a 50-50 solution of concentrated nitric acid and water. In the case of the silicon bronze alloys, it may be necessary to swab the surface to remove a white (SiO_2) deposit.

E2.2.3 Nickel and Nickel-Base Alloys

Material	Formula
Nickel	Nitric Acid or Lepito's Etch
Low-Carbon Nickel	Nitric Acid or Lepito's Etch
Nickel-Copper (400)	Nitric Acid or Lepito's Etch
Nickel-Chromium-Iron (600 & 800)	Aqua Regia or Lepito's Etch

Table E2.2.3
Make up of formulas for aqua regia and lepito's etch

	Aqua Regia (a)	Lepito's Etch (b)
Nitric Acid, Concentrated HNO_3	1 part	3 ml
Hydrochloric Acid, Concentrated HCl	2 parts	10 ml
Ammonium Sulfate $(NH_4)_2 SO_4$	....	1.5 q
Ferric Chloride $FeCl_3$	....	2.5 q
Water	....	7.5 ml

Notes:

1. Warm the parts for faster action.
2. Mix solution as follows:
   (a) Dissolve $(NH_4)_2(SO_4)$ in H_2O.
   (b) Dissolve powdered $FeCl_2$ in warm HCl.
   (c) Mix (1) and (2) and add HNO_3.
3. Etching is accomplished by either swabbing or immersing the specimen.
4. *Titanium.* The following are general purpose etchants which are applied at room temperature by swabbing or immersion of the specimen.

	Kroll's Etch	Keller's Etch
Hydrofluoric Acid (48%)	1 to 3 ml	1/2 ml
Nitric Acid (conc.)	2 to 6 ml	2-1/2 ml
Hydrochloric Acid (conc.)	....	1-1/2 ml
Water	to make 100 ml	to make 100 ml

5. *Zirconium.* Apply the following solution by swab and rinse in cold water.

Hydrofluoric Acid	3 ml
Nitric Acid (conc.)	22 ml
Water	22 ml

E3. Safety Procedures

E3.1 General. All chemicals used as etchants are potentially dangerous. All persons using any of the etchants listed in E2 should be thoroughly familiar with all of the chemicals involved and the proper procedure for handling and mixing these chemicals.

E3.2 Handling and Mixing Acids. Caution must be used in mixing all chemicals, especially strong acids. In all cases, various chemicals should be added slowly INTO the water or solvent while stirring.

E3.2.1 Hydrofluoric Acid, HF. In cases where hydrofluoric acid is used, the solution should be mixed and used in polyethylene vessels. (**Caution:** Hydrofluoric acid or its solutions should not be allowed to contact the skin since it can cause painful, serious ulcers if not washed off immediately.)

E3.3 Basic Recommendations for Handling of Etching Chemicals

E3.3.1 Always use protective garb (gloves, apron, protective glasses or face shield, etc.) when pouring, mixing, or etching.

E3.3.2 Use proper devices (glass or plastic) for weighing, mixing, containing, or storage of solutions.

E3.3.3 Wipe up or flush all spills.

E3.3.4 Dispose of any solutions not properly identified. Do NOT use unidentified solutions; when in doubt, throw it out!

E3.3.5 Store and handle chemicals according to manufacturer's recommendations and observe any printed cautions on chemical containers.

E3.3.6 If not sure about the proper use of a chemical, contact your Safety Department.

Appendix F
Penetrant Examination Procedure

F1. General

This Appendix describes the method which shall be employed when liquid penetrant examination is specified in Section 2 or 3. This procedure is substantially in conformance with ASTM E165, *Recommended Practice for Liquid Penetrant Inspection.* Refer to this Standard for additional details.

F2. Method

A visible solvent removable penetrant shall be used.

F3. Procedure

F3.1 Cleaning. Cleanliness in this application shall be the removal of all rust, scale, welding flux, spatter, grease, oily films, water, dirt, etc., from the surface to be examined.

F3.2 Drying. The surface to be examined shall be allowed to dry for a minimum of 2 minutes prior to application of penetrant.

F3.3 Penetrant Application. The penetrant is applied by brushing or spraying. The penetration time shall be at least 5 minutes. The temperature of the surface to be examined, the liquid penetrant, the cleaner, and the developer shall not be below 40°F (5°C) nor above 125°F (52°C) throughout the examination period.

F3.4 Excess Penetrant Removal. After a minimum of 5 minutes penetration time has elapsed, any penetrant remaining on the surface shall be removed. Penetrants shall be removed from the surface by wiping with rags or cloths moistened with the solvent.

F3.5 After excess penetrant removal, the developer shall be sprayed on the surface to be examined. The developer coating shall be applied lightly and uniformly.

F4. Examination

It is recommended to conduct the examination within 7 minutes after applying the developer. In no case shall the examination be made later than 30 minutes after the developer has been applied.

Metric Conversion Tables

Metric (SI) Equivalents for Section 2 Figures

in.	mm
1/8	3.2
3/8	9.5
1/2	12.7
3/4	19.0
2	50.8
2-1/2	63.6
4-1/2	115
5	127
6	152
7	180
8	203
9	230
18-1/2	470
20	510
30	765

Metric (SI) Equivalents for Section 3 Figures

in.	mm
1/16	1.6
1/8	3.2
3/16	4.8
1/4	6.4
5/16	8.0
3/8	9.5
1/2	12.7
3/4	19.0
1	25.4
1-1/2	38.1
2	50.8

in.	mm
2-1/2	63.6
3	76.2
4	102
4-1/2	115
5	127
6	152
7	180
8	203

Metric (SI) Equivalents for Appendix A Figures

in.	mm
0.003	0.08
0.004	0.10
0.005	0.13
0.010	0.25
0.015	0.38
1/64	0.4
0.20	0.51
0.0299	0.76
1/32	0.79
1/16	1.6
1/8	3.2
0.188	4.78
3/16	4.8
7/32	5.6
0.252	6.40
1/4	6.4
5/16	8.0
11/32	8.7

in.	mm
0.350	8.89
3/8	9.5
15/32	11.9
0.5	12.7
1/2	12.7
0.500	12.70
0.05	12.83
19/32	15.1
5/8	16.0
3/4	19.0
7/8	22.2
1	25.2
1-1/32	26.2
1-1/16	27.0
1-1/8	28.6
1-3/16	30.2
1-1/4	31.7
1-3/8	34.9
1-1/2	38.1
1-11/16	42.9
1-15/16	47.6
2	50.8
2-1/16	52.4
2-1/4	57.2
2-1/2	63.6
3	76.2
3-3/8	85.7
3-7/8	98.4
4-1/2	115
6	152
6-3/8	171
7-1/2	190
9	230
10	255